Critters of Washington

Pocket Guide to Animals in Your State

ALEX TROUTMAN

produced in cooperation with
Wildlife Forever

About Wildlife Forever

Wildlife Forever works to conserve America's outdoor heritage through conservation education, preservation of habitat, and scientific management of fish and wildlife. Wildlife Forever is a 501c3 nonprofit organization dedicated to restoring habitat and teaching the next generation about conservation. Become a member and learn more about innovative programs like the Art of Conservation®, The Fish and Songbird Art Contests®, Clean Drain Dry Initiative™, and Prairie City USA®. For more information, visit wildlifeforever.org.

Thank you to Ann McCarthy, the original creator of the Critters series, for her dedication to wildlife conservation and to environmental education. Ann dedicates her work to her daughters, Megan and Katharine Anderson.

Front cover photos by **Antonio Guillem/shutterstock.com:** grizzly bear; **Jason Mintzer/shutterstock.com:** California mountain kingsnake; **Alan B. Schroeder /shutterstock.com:** American robin
Back cover photo by **Wirestock Creators/Shutterstock.com:** orca

Edited by Brett Ortler and Jenna Barron
Cover and book design by Jonathan Norberg
Proofreader: Emily Beaumont

10 9 8 7 6 5 4 3 2 1

Critters of Washington

The first *Critters* books were produced by Wildlife Forever. AdventureKEEN is grateful for its continued partnership and advocacy on behalf of the natural world.

Published by Adventure Publications
An imprint of AdventureKEEN
310 Garfield Street South, Cambridge, Minnesota 55008
(800) 678-7006
www.adventurepublications.net

Printed in China
Cataloging-in-Publication data is available from the Library of Congress
ISBN 978-1-64755-501-6 (pbk.); 978-1-64755-502-3 (ebook)

Acknowledgments

I want to thank everyone who believed in and supported me over the years—a host of friends, family, and teachers. I want to especially thank my mom and my siblings Van, Bre, and TJ.

Dedication

I dedicate this book to my brother Van:
May you continue to enjoy the birds and wildlife in heaven.

This book is for all the kids who have a passion for nature and the outdoors, especially ones who identify as Black, Brown, Indigenous, and People of Color. May this be an encouragement to never give up. And if you have a dream and passion for something, pursue it relentlessly. I also hope to set an example that you can be successful as your full, authentic self!

Lastly, I dedicate this book to all those with ADHD and dyslexia, as well as all other members of the neurodivergent community. While our quirks make things more challenging, our goals are not impossible to reach; sometimes it takes a little more time and help, but we, too, can succeed!

Contents

Mammals

Birds

Reptiles and Amphibians

Introduction

My passion for nature started when I was young. I was always amazed by the sunlit fiery glow of the red-tailed hawks as they soared overhead when I went fishing with my family. The red-tailed hawk was my spark bird—the bird that captures your attention and gets you into birding. Through my many encounters with red-tailed hawks, and other species like garter snakes and coyotes, I found a passion for nature and the environment. Stumbling across conservationists like Steve Irwin, Jeff Corwin, and Jack Hanna introduced me to the field of Wildlife Biology as a career and gave birth to a dream that I was able to accomplish and live out: serving as a Fish and Wildlife Biologist for governmental agencies, as well as in the private sector.

My childhood dream was driven by a desire to learn more about the different types of ecosystems and the animals that call our wild places home. Books and field guides like this one whet my thirst for knowledge. Even before I could fully understand the words on the pages, I was drawn to books and flashcards that had animals on them. I could soon identify every animal I was shown and tell a fact about it. I hope that this edition of *Critters of Washington* can be the fuel that sustains your passion for not only learning about wildlife, but also for caring for the environment and making sure that all are welcome in the outdoors. For others, may this book be the spark that ignites a flame for wildlife preservation and environmental stewardship. I hope that this book inspires children from lower socioeconomic and minority backgrounds to pursue their dreams to the fullest and be unapologetically themselves.

By profession, I'm a Fish and Wildlife Biologist, and I'm a nature enthusiast through and through. My love for nature includes making sure that everyone has an equal opportunity to enjoy the outdoors in their own way. So, as you use this book, I encourage you to be intentional in inviting others to appreciate nature with you. Enjoy your discoveries and stay curious!

–Alex Troutman

Washington: The Evergreen State

Washington is known for its fantastic national parks; big cities like Seattle; and its production of fruits like apples, cherries, and pears. It is called the Evergreen State because of its vast forests, which supply much of the lumber in the country as well as homes for many wild critters. The first inhabitants of the area actually traveled from Asia using the Bering Strait, a strip of land that once connected the two continents 20,000 years ago. Many Indigenous tribes lived on the land for thousands of years before Lewis and Clark explored the West in 1805. The area belonged to both the British and Americans until 1846 and officially became a state in 1889.

Washington borders Canada and the Pacific Ocean, as well as Oregon to the south and Idaho to the east. Eastern Washington is made up of dry areas called shrub-steppe, which are perfect homes for animals like rubber boas and badgers. In the northwest corner of the state are the Olympic Mountains, with wet forests that are part of a national park. The Cascade Range makes up a strip of western Washington, where the larger mountains are, such as Mount St. Helens and Mount Rainier. The Pacific Coast is where you can see marine life like orcas and humpback whales. The northernmost point of the Rocky Mountains reaches Washington in the southeast. Across the southern part of the state is the Columbia Plateau, where lava formed dry canyons with clay soil called scablands.

These environments are home to many animals, including 141 species of mammals, 341 species of birds, and at least 53 species of reptiles and amphibians, not to mention fish, countless insects and spiders, plants, and more. This is your guide to the animals, birds, reptiles, and amphibians that call Washington home.

Some of Washington's most iconic plants, animals, and other natural resources are now officially recognized as state symbols. Get to know them below and see if you can spot them all! You'll probably encounter the state nickname and motto, so I've included them here too.

State Bird: willow goldfinch

State Fossil: Columbian mammoth

State Tree: western hemlock

State Flower: coast rhododendron

State Fish: steelhead trout

State Amphibian: Pacific chorus frog

State Fruit: apple

State Marine Mammal: orca

State Nickname: The Evergreen State

State Insect: green darner dragonfly

State Motto: *"Into the Future"*

How to Use This Guide

This book is your introduction to some of the wonderful critters found in Washington; it includes 26 mammals, 27 birds, and 15 reptiles and amphibians. It includes some animals you probably already know, such as black bears and bald eagles, but others you may not know about, such as coastal tailed frogs or varied thrushes. I've selected the species in this book because they are widespread (northern raccoon, page 44), abundant (red-winged blackbird, page 96), or well-known but best observed from a safe distance (rubber boa, page 126).

The book is organized by types of animals: mammals, birds, and reptiles and amphibians. Within each section, the animals are in alphabetical order. If you'd like to look for a critter quickly, turn to the checklist (page 140), which you can also use to keep track of how many animals you've seen! For each species, you'll see a photo of the animal, along with neat facts and information on the animal's habitat, diet, its predators, how it raises its young, and more.

Safety Note

Nature can be unpredictable, so don't go outdoors alone, and always tell an adult when you're going outside. All wild animals should be treated with respect. If you see one—big or small—don't get close to it or attempt to touch or feed it. Instead, keep your distance and enjoy spotting it. If you can, snap some pictures with a camera or make a quick drawing using a sketchbook. If the animal is getting too close, is acting strangely, or seems sick or injured, tell an adult right away, as it might have rabies, a disease that can affect mammals. The good news is there's a rabies vaccine, so it's important to visit a doctor right away if you get bit or scratched by a wild animal.

Notes About Icons

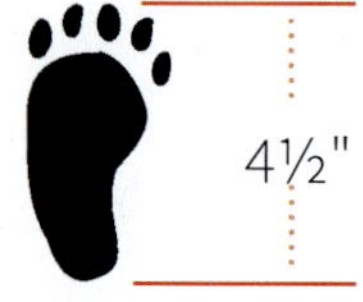

Each species page includes basic information about an animal, from what it eats to how it survives the winter. The book also includes information that's neat to know; in the mammals section, each page includes a simple track illustration of the animal, with approximate track size included. And along the bottom, there is an example track pattern for the mammal, with the exception for those that primarily fly (such as bats).

On the left-hand page for each mammal, a rough-size illustration is included that shows how big the animal is when compared to a basketball.

Also on the left-hand page, there are icons that tell you when each animal is most active: nocturnal (at night), diurnal (during the day), or crepuscular (at dawn/dusk), so you know when to look. If an animal has a "zzz" icon, it hibernates during the winter. Some animals hibernate every winter, and their internal processes (breathing and heartbeat) slow down almost entirely. Other animals only partially hibernate, but this still helps them save energy and survive through the coldest part of the year.

nocturnal
(active at night)

diurnal
(active during day)

crepuscular
(most active at dawn and dusk)

hibernates/deep sleeper
(dormant during winter)

ground nest

cup nest

platform nest

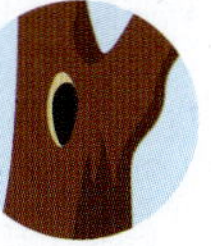

cavity nest

migrates

On the left-hand side of each bird page, the nest for the species is shown, along with information on whether or not the bird migrates; on the right-hand side, there's information on where it goes.

Did you know?
Badgers are solitary animals, but they will sometimes hunt with coyotes in a team. A coyote will chase prey into the badger's den, and the badger will chase or dig out the prey that coyotes like. When a badger is threatened, it will back into its burrow and show its teeth.

Size Comparison

Most Active

Track Size

Hibernates

American Badger

Taxidea taxus

Size: 2–3 feet long; weighs 8–25 pounds

Habitat: Savannas, grasslands, and meadows

Range: Can be found throughout the Midwest and westward through the Great Plains to the western coast of the United States and southward into Mexico. In Washington, they can be found in the eastern portion of the state.

Food: Carnivores; they eat pocket gophers, moles, ground squirrels, and other rodents. They will also eat dead animals (or carrion), fish, reptiles, and a few types of birds, especially ground-nesting birds.

Den: Badgers are fossorial (a digging animal that spends a lot of time underground); they build many dens or burrows throughout their range. Most dens are used to store food, but badgers also use dens to sleep in and raise their young. Dens can be over 10 feet deep and 4 feet wide.

Young: Cubs are born, with eyes closed, usually in April or May in litters of 2–3. Extensive care is provided by the mom for up to 3 months. After another 2–3 months, the young will gain their independence.

Predators: Bears, bobcats, mountain lions, coyotes, golden eagles, and humans

Tracks: The front tracks are 2¾ inches long and 2 inches wide.

The American badger is a short, bulky mammal with grayish-to-dirty-red fur. Badgers have a distinctive face with a series of cream-and-white stripes offset by a black background.

Did you know?

Beavers are rodents! Yes, these flat-tailed mammals are rodents, like rats and squirrels. In fact, they are the largest native rodents in North America. Just like other rodents, beavers have large incisors, which they use to chew through trees to build dams and dens. Beavers are the original wetland engineers. By damming rivers and streams, beavers create ponds and wetlands.

Size Comparison

Most Active

Track Size

American Beaver

Castor canadensis

Size: Body is 25–30 inches long; tail is 9–13 inches long; weighs 30–70 pounds

Habitat: Wooded wetland areas near ponds, streams, and lakes

Range: Beavers can be found throughout Washington; found in much of the US.

Food: Leaves, twigs, and stems; they also feed on fruits and aquatic plant roots. Throughout the year they gather and store tree cuttings, which they eat in winter.

Den: A beaver's home is called a lodge. It consists of a pile of branches that is splattered with mud and vegetation. Lodges are constructed on the banks of lakes and streams and have exits and entrances that are underwater.

Young: Young beavers (kits) are born in late April through May and June in litters of 3–4. After two years they are considered mature and will be forced out of the den.

Predators: Bobcats, mountain lions, bears, wolves, and coyotes. Human trappers are major predators too.

Tracks: A beaver's front foot looks a lot like your hand; it has five fingers. The hind (back) foot is long, with five separate toes that have webbing or extra skin between them.

Beavers range from dark brown to reddish brown. They have a stocky body with hind legs that are longer than the front legs. The beaver's body is covered in dense fur, but its tail is naked and has special blood vessels that help it cool or warm its body.

Did you know?

Female bears weigh between 90 and 300 pounds and are smaller than the average adult human male in the US. But don't let their small size fool you; with a bite force around 800 pounds per square inch (PSI) and swiping force of over 400 pounds, these bears are not to be taken lightly.

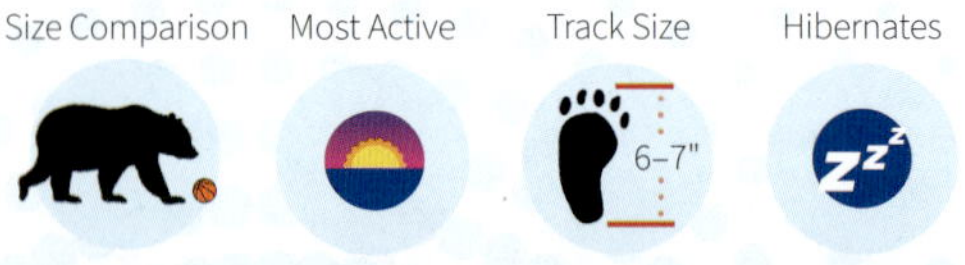

Size Comparison

Most Active

Track Size

Hibernates

Black Bear

Ursus americanus

Size: 5–6 feet long (nose to tail); weighs 90–600 pounds

Habitat: Forests, lowland areas, and swamps

Range: Black bears can be found in many parts of North America, from Alaska down through Canada and into Mexico. In Washington, they are found throughout much of the state, minus the area of the Interior Columbia Basin.

Food: Berries, fish, seeded crops, small mammals, wild grapes, tree shoots, ants, bees, beavers, and even deer fawns

Den: Denning usually starts in December, with bears emerging in late March or April. Dens can be either dug (out of a hillside, for example) or constructed with materials such as leaves, grass, and moss.

Young: Two cubs are usually born at one time (a litter), often in January. Cubs are born without fur and blind, with pink skin. They weigh 8–16 ounces.

Predators: Humans and other bears. Sometimes, other carnivores, such as mountain lions, wolves, coyotes, or even bobcats, will prey on black bears. Cubs are especially vulnerable.

Tracks: Front print is usually 4–6 inches long and 3½–5 inches wide, with the hind foot being 6–7 inches long and 3½–5 inches wide. The feet have five toes.

Black bears are usually black in color, but they can be many different variations of black and brown. Some even have grayish, reddish, or blonde fur.

Did you know?

Bobcats get their name from their short tail; a "bob" is a type of short haircut. They have the largest range of all wild cats in the United States. Bobcats can even hunt prey much larger than themselves; in fact, they can take down prey that is over four times their size, such as white-tailed deer!

Size Comparison

Most Active

Track Size

Bobcat

Lynx rufus

Size: 27–48 inches head to tail; males weigh around 30 pounds, while females weigh 24 pounds or so.

Habitat: Dense forests, scrub areas (forests of low trees and bushes), swamps, and even some urban (city) areas

Range: They are widespread throughout the United States. Bobcats are found in the mainland of Washington.

Food: Squirrels, birds, rabbits, snowshoe hares, and white-tailed deer fawns; occasionally even adult deer!

Den: Dense shrubs, caves, or even hollow trees; dens can be lined with leaves or moss.

Young: Bobcats usually breed in early winter through spring. Females give birth to a litter of 2–4 kittens. Bobcats become independent around 7–8 months, and they reach reproductive maturity at 1 year for females and at 2 years for males.

Predators: Occasionally fishers and coyotes; humans also hunt and trap bobcats for fur.

Tracks: Roughly 2 inches wide; both front and back paws have four toe pads and a carpal pad (a pad below the toe pads).

Bobcats have a white belly and a brown or pale-gray top with black spots. The tail usually has a black tip. They are mostly crepuscular (say it, cre-pus-cue-lar), which means they are most active in the dawn and twilight hours.

Did you know?

California gray whales have one of the longest migrations of any mammal, covering distances of more than 10,000 miles roundtrip. The amount of barnacles and whale lice on gray whales' bodies can weigh over 400 pounds!

Size Comparison

Most Active

California Gray Whale

Eschrichtius robustus

Size: 40–49 feet long; weighs 90,000 pounds or 45 tons

Habitat: Shallow coastal waters in the northern Pacific Ocean

Range: In the summer, they can be found in the feeding grounds along the Pacific Coast of Alaska, British Columbia, Washington, Oregon, and California. In the fall, they migrate south to wintering and calving areas off the coast of Baja California, Mexico.

Food: They feed mainly on small marine invertebrates like marine worms, crustaceans, and mollusks.

Den: No den; calves are born in shallow lagoons and bays.

Young: A single calf is born after a year-long pregnancy. At birth, calves are around 14 feet long and weigh 2,000 pounds. They reach reproductive maturity between 6 and 12 years and can live up to 75 years or more.

Predators: Orcas

Tracks: Fully aquatic mammal with fins instead of feet

California gray whales can be over 40 feet long and weigh 90,000 pounds. Females are a little bit bigger than males. Gray whales have small eyes, a gray body, and broad flippers. They do not have a dorsal (back) fin; instead, they have a hump followed by several smaller bumps that are known as knuckles. They have a tail fluke that is about 10 feet wide. Calves are a darker gray and wrinkly at birth and will lighten as they age. Like many whale species, they have patches of whale lice and barnacles on their body.

This animal is fully aquatic, so it does not leave tracks.

Did you know?

San Miguel Island, off the coast of California, is the largest rookery (gathering) of breeding California sea lions. There can be over 70,000 animals on the island. Pups have distinct whines and noises that females are able to recognize even from hundreds of feet away. Sea lions use a technique called rafting in order to rest and regulate body temperature; they hold their flippers above the water while their head is in the water.

Size Comparison

Most Active

Individual tracks of these marine mammals are rare.

California Sea Lion

Zalophus californianus

Size: Male: up to 7½ feet long; weighs up to 700 pounds. Female: up to 6 feet long; weighs 240 pounds.

Habitat: Open ocean, coastal waters, coves, harbors, rocky shores, and marinas; they will rest and sleep on human-made objects like docks.

Range: They can be found in North America from Canada to the southern tip of Baja California, Mexico. They are found year-round on the coast of Washington.

Food: They eat several species of fish, crabs, and squids.

Den: Does not den; pups are left on shore or rocky areas as mom is out fishing for food.

Young: Pups are born in June and July. At birth, they are 13–25 pounds (sometimes even over 40). They nurse for about 6 months, and they will stay with their mom for a year as she teaches them to hunt and swim. They reach reproductive maturity at 3–8 years old for females and 6–10 years old for males.

Predators: Sharks and orcas; as pups: eagles and coyotes

Tracks: A mostly aquatic mammal, its tracks are in the shape of its dragging flippers.

Adult females and juveniles are shaped like a 2-liter bottle and are smaller than the males. They are light brown to tan in color. Adult males are larger and dark brown to almost black in color. At birth, pups are dark brown and will shed their birth coats at around 4 months old. Sea lions have a long snout, visible ear flaps, and strong wide front and hind flippers that they are able to rotate underneath their bodies to help them move on land.

Did you know?

At one time, coyotes were only found in the central and western parts of the US, but now, with the help of humans (eliminating predators and clearing forests), they can be found throughout most of the country.

Size Comparison

Most Active

Track Size

Coyote

Canis latrans

Size: 3–4 feet long; weighs 21–50 pounds

Habitat: Urban and suburban areas, woodlands, grasslands, and farm fields

Range: In Washington, coyotes are found statewide. They are also found throughout the US and Mexico, the northern parts of Central America, and in southern Canada.

Food: A variety of prey, including rodents, birds, deer, and sometimes livestock

Den: Coyotes will dig their own dens but will often use old fox or badger dens or hollow logs.

Young: 5–7 pups, independent around 8–10 months

Predators: Bears and wolves; humans trap and kill for pelts and to "protect" livestock.

Tracks: Four toes and a carpal pad (the single pad below the toe pads) can be seen on all four feet.

Coyotes have brown, reddish-brown, or gray back fur with a lighter gray-to-white belly. They have a longer muzzle than other wild canines. They are active mostly during the night (nocturnal) but also during the twilight and dawn hours (crepuscular).

Did you know?

All wolves in the United States (except the red wolf in the Southeast) are gray wolves! Each area has its own subspecies (a group that is a little different physically or genetically), and their common names are often based on their habitat. Gray wolves are often referred to as timber wolves. Wolves were believed to be extinct in Washington until 2008, when a breeding pair was documented in the state for the first time in over 70 years.

Size Comparison

Most Active

Track Size

Gray Wolf

Canis lupus

Size: 5–6½ feet long; weighs 60–130 pounds

Habitat: Forests and grasslands

Range: They can be found throughout much of Canada and Alaska, as well as many states out West and in a few states of northern New England. In Washington, they can be found in the northern Cascade Mountains and the northeastern corner of the state, although they were once common throughout the state.

Food: Deer; moose; elk; and smaller animals like rabbits, beavers, and birds

Den: For the first couple of weeks, the female stays with the pups to keep them warm and fed. During this time, she is totally dependent on the remainder of the pack to provide her with food. Once the pups are large enough to be alone, the female can leave them and hunt to support the growing pack.

Young: Pups are born in April or May; 4 to 7 pups are born at one time; they stay in the den for 6–8 weeks and eventually leave the pack at 1–2 years of age.

Predators: Bears, other wolves, coyotes, and humans

Tracks: Front paws are around 5 inches long; hind paws are 4 inches long; both front and hind paws have a width of 3–3½ inches.

Gray wolves can be gray, black, or even red. Wolves live in groups called packs. Packs can be as small as 2 wolves and as large as 13 or more. Wolves will hunt in packs; when hunts are successful, the whole pack will feed on the kill.

Did you know?

Grizzly bears can run up to 35 miles per hour in short bursts. Their roar can be heard from as far as 1 mile away. They are important to the ecosystem in many ways. They spread seeds through their waste after eating fruit. They also help to keep the herbivore population down, a system that, if left unchecked, could lead to overgrazing of plants by these species.

Size Comparison

Most Active

Track Size

Hibernates

Grizzly Bear

Ursus arctos horribilis

Size: 78–95 inches long; shoulder height is 40 inches; weighs 350–700 pounds

Habitat: Mountainous areas, forests, woodlands, alpine meadows, prairies, and areas near rivers and streams

Range: They can be found from Alaska down into Canada and Washington and as far east as Montana. They can be found in the Selkirk Mountains of northeastern Washington and in two counties on the Canadian border.

Food: Omnivores, they feed on berries, fruits, and roots, as well as fish, small mammals, elk, and deer.

Den: They will dig a den in the ground or under a rock, or use a rock crevice, hollow tree, or cave, usually on a north-facing hillside.

Young: Two cubs are born per year in winter. They feed on milk until summer and grow rapidly, going from 9 to 99 pounds in 2 years. They reach reproductive maturity at 5 years.

Predators: Humans

Tracks: Front foot is stubby compared to the back foot. Both feet have five toes and a tarsal pad. Front claws are 2–4 inches long. Front track is 7 inches long and 5 inches wide, and the hind foot is 11 inches long and 6 inches wide.

Grizzly bears are large with dark- to light-brown (or even golden-to-yellowish) fur. They have a shoulder hump, thick body, and large feet with long claws. They have a circular face with a long nose; brown eyes; and short, rounded ears.

Did you know?

Harbor seals are one of the most common marine mammals to be seen on both the Atlantic and Pacific Coasts! Due to their pelvic bones being fused, they are unable to move their hind (back) flippers to walk. Instead, they use caterpillar-like movements to cross land; this is called galumphing.

Size Comparison

Most Active

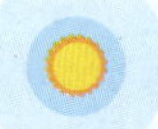

Individual tracks of these marine mammals are rare.

Harbor Seal

Phoca vitulina

Size: 5–6 feet long; weighs 90–350 pounds (or more for males)

Habitat: Coastal areas such as sandy beaches; tidal mudflats; sand bars; and exposed, rocky shorelines

Range: They are found along the entire eastern coast to Florida and down the western coast of the United States. In Washington, they are found throughout the western coastline waters.

Food: Fish, shellfish, and crustaceans

Den: Aquatic mammals do not den.

Young: Pups weighing about 24 pounds are born around 10 months after breeding and can swim minutes after being born. They drink their mother's milk for up to 6 weeks and reach reproductive maturity at 3–7 years.

Predators: Orcas; great white and other shark species; Steller sea lions and eagles

Tracks: They have webbed front feet with five toes, or digits, on them. Hind tracks are rarely left due to limited movement on land.

Harbor seals are brown, gray, or variations of browns and grays; some even have splotches or rings on them. They have stout, round bodies with short flippers. They have big heads with nostrils that are chevron- or V-shaped. They do not have external ear flaps like similar species, such as sea lions.

Did you know?

The hoary bat is one of the most widespread bats in North and South America. They are the only bat species found in Hawaii. They can fly 13 miles per hour and as high as 8,000 feet during their long-distance migrations.

Size Comparison

Most Active

Hoary Bat

Lasiurus cinereus

Size: 5⅛–6 inches long; wingspan of 17 inches; weighs ¾–1¼ ounces

Habitat: Dunes, savannas or grasslands, forests, and deserts

Range: They are found from Canada down through the United States, through Mexico and Central America, and into South America. In Washington, they are widespread across the state.

Food: They are carnivores that feed mainly on moths but will also eat flies, beetles, and smaller wasp species.

Den: They roost in the foliage or leaves of trees and are mostly solitary roosters, unlike many other bat species.

Young: Young are born while mom is hanging upside down. They have brownish ears and eyes that are closed, and they have fine silver hair. Their eyes and ears open at around day 3. They hang on their mom during the day; at night, they hang onto tree leaves while mom is out finding food. They learn to fly at around day 33.

Predators: Hawks, owls, and snakes

Tracks: Though they are rarely on the ground to leave a track, it would show one thumbprint on the forearm and the hind footprint.

Hoary bats have thick, dark fur with a whitish or frosty-hued upper tip of each hair, which is how they get their name. They have a pale-yellow throat patch, small eyes, and rounded ears and noses.

The hoary bat does not often leave tracks.

Did you know?

The humpback whale gets its name from the large hump on its back. Humpback whales come in various color patterns, and the spots and scars on their tail fluke can help researchers identify individuals. Its front fins can be over 15 feet long, up to a third of its total length.

Size Comparison

Most Active

Humpback Whale

Megaptera novaeangliae

Size: 48–63 feet long; weighs 80,000 pounds, or 40 tons

Habitat: Open water; visitors to bays and harbors

Range: Can be found in every ocean. They can be seen off the coast of Washington, with most sightings occurring from July to September.

Food: They are carnivores that mainly eat krill and fish.

Den: No den; gives live birth in the ocean

Young: After 11½ months of pregnancy, females usually give birth to one calf that is over 10 feet long and weighs over 1,000 pounds. After 5–7 months, calves are weaned and will remain with their mom for a year. Humpback whales are reproductively mature at around 4–5 years old.

Predators: Humans (boat strikes), orcas, and large sharks

Tracks: Fully aquatic animals with flippers; no tracks are left.

Humpback whales sport a black upper body and white underside. They have over 20 grooves running down their throat and underside. Humpback whales are stout and have long flippers and a short dorsal fin. Flippers and fluke (tail) are jagged.

This animal is fully aquatic, so it does not leave tracks.

Did you know?

Mountain goats are not actually goats; they are in the same family as antelopes, gazelles, and cows in the Bovidae family. Mountain goat males are called billies, and the females are called nannies. Mountain goats can move each side of their hooves independently.

Size Comparison

Most Active

Track Size

Mountain Goat

Oreamnos americanus

Size: 49–70½ inches long; weighs 125–180 pounds

Habitat: Mountains, rocky cliffs, bluffs, and alpine areas

Range: They can be found in western North America from Alaska to Colorado. In Washington, they are mostly found near the Cascade Range. There are smaller populations in the southeastern areas of the state and near the Idaho border.

Food: They are herbivores that graze on grasses, shrubs, and flowers. During winter, they feed on berries, lichen, and young trees.

Den: No den; they give birth along a mountain or hillside.

Young: One kid is born in May after a 180-day pregnancy. Young are about 4 pounds and can walk shortly after being born. Kids will stay with their mother until the next breeding season or longer. They reach reproductive maturity at around 4–5 years.

Predators: Bears, wolves, golden eagles, and mountain lions

Tracks: Tracks are two marks on the ground that are 3–4 inches long and 2–3 inches wide.

Mountain goats have short, shiny black horns. During the winter, they have long, shaggy white fur. Males have horns that are wide at the base and more evenly arched, while females have horns that are more curved at the tip. Males are larger than females.

Did you know?

Mountain lions are the second-largest cat in the western part of the world. The largest is the jaguar. Mountain lions do not roar like other big cats, but rather they scream! They also make other sounds similar to pet cats, like hissing and purring. Mountain lions can jump as high as 18 feet off the ground into a tree.

Size Comparison

Most Active

Track Size

Mountain Lion

Puma concolor

Size: 6–8 feet long; weighs 100–154 pounds

Habitat: Grasslands, deserts, wetlands, shrublands, forests, swamps, and upland forests

Range: They can be seen from northern Canada to Argentina. They are widespread throughout Washington.

Food: Deer, wild boars, raccoons, birds, rabbits, mice, and occasionally livestock

Den: Will den in caves, rock piles and crevices, and thickets. Dens are usually lined with plants.

Young: 1–6 kittens are born with spots almost 100 days after mating. Weaning takes place around day 4, and the young kits will stay with their mom another year or two. Spots fade at around 6 months. Males reach reproductive maturity at around 3 years old and females around 2½ years, though they usually do not reproduce until they have a permanent home territory.

Predators: No natural predators, but they will sometimes get in territory disputes with other large carnivores.

Tracks: Front tracks are 3¼ inches long and wide. The back or hind tracks are 3 inches long and wide.

Mountain lion fur is golden tan to dusky brown on the back; its underside is a pale buff color with a white throat and chest area. They have a pink nose, black ear tips, and a smoky gray-black muzzle. The tip of their tail is black like their ears, and their eyes are brown. Their tail is long and makes up a third of their body length. Kittens have spots and smoky-blue eyes.

Did you know?

Mule deer get their name from their large ears that resemble those of mules. When they first emerge, a deer's antlers are covered in a special skin called velvet. Deer can run up to 40 miles per hour and can jump over 8 feet vertically (high) and over 15 feet horizontally (long). Mule deer and other deer species do not have teeth on the top of their mouth, just a hard palate.

Size Comparison

Most Active

Track Size

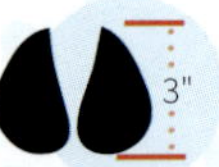

Mule Deer

Odocoileus hemionus

Size: 4–7 feet long; 3–3½ feet tall; weighs 200–260 pounds

Habitat: Forest edges, brushy fields, woody farmlands, prairies, deserts, and mountainous and rocky areas

Range: They are found from Alaska to as far south as northern Mexico and as far east as Nebraska. In Washington, they are found throughout the state.

Food: Fruits, grasses, trees, shrubs, nuts, and bark

Den: Deer do not den, but they will bed down in tall grasses and shrub areas.

Young: After a 200-day pregnancy, fawns (young) are born with spots and weigh around 5½ pounds. They rest in the same spot where they're born for about a week or so. They lose their spots and are weaned at around 2–2½ months. Females reach reproductive maturity at around 18 months, and males reach maturity a little earlier but will not mate until they can participate in the rut (fight for mates) at age 3 or 4.

Predators: Wolves, coyotes, bears, bobcats, mountain lions, and humans

Tracks: Both front and hind feet have two teardrop- or comma-shaped toes.

Mule deer have large ears and are overall brownish gray with a white patch of fur on their rump and a black-tipped white tail.

Did you know?

A single porcupine can have over 30,000 quills that it can use to protect itself from predators. Porcupine quills are hollow and can be over 2 inches long. Porcupines are herbivores (plant eaters), and they have a special bacteria in their digestive system to help them break down the plant material.

Size Comparison

Most Active

Track Size

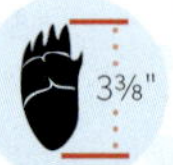

North American Porcupine

Erethizon dorsatum

Size: 2–3 feet long; weighs 10–25 pounds

Habitat: Forested areas, grasslands, and deserts

Range: They can be found in most of Washington. They are also found throughout Canada and various areas across the northern and western US.

Food: Skunk cabbage, clovers, twigs, leaves, and tree bark

Den: They den in hollow logs and tree cavities.

Young: One young (porcupette) is born between May and July; they weigh a pound at birth and have 1-inch quills.

Predators: Lynx, bobcats, coyotes, owls, and fishers

Tracks: The front foot is shorter than the hind foot; the hind foot has five toes, while the front only has four toes.

The North American porcupine is mostly nocturnal; its fur is black to gray and shades of brown with obvious quills. When threatened, it will turn around and strike an attacker with its tail quills, which are 4 inches long and are like needles. With that many quills, porcupines sometimes accidentally poke themselves. To protect itself (from itself), the porcupine has a special substance on its quills that acts like an antibiotic (or medicine). This prevents it from getting infected after an accidental poke. Like other mammals, porcupines need salt and will chew on, or lick, objects with the mineral to fulfill that craving. Sometimes this leads to porcupines chewing on human-made structures.

Did you know?

The raccoon is great at catching fish and other aquatic animals, such as mussels and crayfish. They are also excellent swimmers, but they apparently avoid swimming because the water makes their fur heavy. Raccoons can turn their feet 180 degrees; this helps them when climbing, especially when going headfirst down trees.

Size Comparison

Most Active

Track Size

Hibernates

Northern Raccoon

Procyon lotor

Size: 24–40 inches long; weighs 15–28 pounds

Habitat: Woody areas, grasslands, suburban and urban areas, wetlands, and marshes

Range: They are found throughout the US; they are also found in Mexico and southern Canada. They are widespread in Washington.

Food: Eggs, insects, garbage, garden plants, berries, nuts, fish, carrion, small mammals, and aquatic invertebrates like crayfish and mussels

Den: Raccoon dens are built in hollow trees, abandoned burrows, caves, and human-made structures.

Young: 2–6 young (kits) are born around March through July. They are born weighing 2 ounces, are around 4 inches long, and are blind with lightly colored fur.

Predators: Coyotes, foxes, bobcats, humans, and even large birds of prey

Tracks: Their front tracks resemble human handprints. The back tracks sort of look like human footprints.

The northern raccoon has dense fur with variations of brown, black, and white streaks. It has black, mask-like markings on its face and a black-and-gray/brownish ringed tail. During the fall, it will grow a thick layer of fat to stay warm in the winter.

Did you know?

Otters are good swimmers and can close their nostrils while diving. This allows them to dive for as long as 8 minutes and to depths of over 50 feet. Otter fur is the thickest of all mammal fur. River otters have an incredible 67,000 hairs for every square centimeter!

Size Comparison

Most Active

Track Size

Northern River Otter

Lontra canadensis

Size: 29–48 inches long; weighs 10–33 pounds

Habitat: Lakes, marshes, rivers, and large streams; suburban areas

Range: They are widespread throughout Washington. They are also found across much of the US, except parts of the Southwest and portions of the central US.

Food: Fish, frogs, snakes, crabs, crawfish, mussels, birds, eggs, turtles, and small mammals. They sometimes eat aquatic vegetation too.

Den: They den in burrows along the river, usually under rocks, riverbanks, hollow trees, and vegetation.

Young: 2–4 young (pups) are born between November and May. Pups are born with their eyes closed. They will leave the area at around 6 months old and reach full maturity at around 2 or 3 years.

Predators: Coyotes, bobcats, bears, and dogs

Tracks: Their feet have nonretractable claws and are webbed.

Northern river otters have thick, dark-brown fur and a long, slender body. Their fur is made up of two types: a short under-coat and a coarse top coat that repels water. They have webbed feet and a layer of fat that helps keep them warm in cold water.

Did you know?

The northern sea otter is the smallest marine mammal in North America, but it's the largest member of the Weasel family! Yes, sea otters are related to ferrets and mink. Sea otters have the densest fur of all mammals. Their fur has 1 million hairs per square inch.

Size Comparison

Most Active

Northern Sea Otter

Enhydra lutris kenyoni

Size: 4 feet long; weighs 70 pounds

Habitat: Kelp forests and estuary habitats (where the ocean meets a river or stream)

Range: Historically, sea otters could be found from the coast of Alaska to Mexico. Today, the northern subspecies can be found off the coast of Alaska, Canada, and Washington. In Washington, they can be found from Pillar Point in the Strait of Juan de Fuca southward to the water south of Destruction Island.

Food: Crabs, clams, mussels, snails, and abalones (mollusks)

Den: No den is made; births take place in water.

Young: Pups are born in the water and weigh about 3–5 pounds. Females carry their babies on their bellies while nursing. Pups are weaned at about 6 months. When not carrying her young, the mom will wrap kelp around them, so they won't drift away. At around 2 months old, they will start to dive. They are mature at around 4 years for females and 5–6 years for males.

Predators: Orcas, coyotes, sharks, and eagles

Tracks: They are aquatic mammals that rarely travel on land. Both front and hind feet have 5 digits, and the back feet are webbed.

Sea otters' fur is made of two layers; the first layer is undercoat, and the second is made of long guard hairs that help them stay dry by trapping air against the skin. They have furry faces, round eyes and ears, long whiskers, and a short nose. They have short front legs with retractable claws and long back legs.

Sea otters rarely go on land to leave tracks.

Did you know?
The marmot is the largest ground squirrel found in Washington and the largest marmot species in North America. Olympic marmots are endemic (or only found) in Olympic National Park. They spend about 80% of their time in burrows that may be as deep as 16 feet or more.

Size Comparison

Most Active

Track Size

Hibernates

Olympic Marmot

Marmota olympus

Size: 26–30 inches long; weighs 7–24 pounds

Habitat: Alpine meadows, fields, pastures, and rocky areas

Range: They are only found in the Olympic Peninsula, which is mostly in Olympic National Park in Washington.

Food: They are herbivores that eat flower stalks, grasses, flowers, and seeds.

Den: Dens consist of an underground burrow with complex internal structures and multiple nesting areas. Nesting chambers are usually away from activity, toward the end of the burrow, and lined with grasses.

Young: 30 days after breeding, a litter of 3–5 pups is born. Pups will leave the nest at about 3 weeks old and are fully weaned at around week 10. Males are driven from the nest, while females will stay in the group. They reach maturity at around 2 years old but will not mate until year 3.

Predators: Coyotes, hawks, mountain lions, bears, owls, bobcats, golden eagles, and badgers

Tracks: They have four toes on the front feet and five on the hind feet. Front feet are 1½–2½ inches long and 1–1½ inches wide. Hind feet are 2–2½ inches long and wide.

Olympic marmots are brown, tan, or sometimes yellowish in color during the spring and dark brown to black in the fall. They have a dark head with small ears and a short muzzle. They have white patches on the muzzle and between their eyes. They gain two black shoulder patches in the summer.

Did you know?

Orcas, or killer whales, are not actually whales but dolphins! In fact, they are the largest species of dolphin in the world. Males can have a dorsal fin that is 6 feet tall, which is larger than the average height for a male human. Of all cetaceans (the group composed of whales and dolphins), they have the widest-spread range.

Size Comparison

Most Active

Orca

Orcinus orca

Size: 23–32 feet long; weighs 22,000 pounds or 11 tons

Habitat: Open seas and coastal waters

Range: Found in every ocean in the world. In Washington, they can be found in most of the coastal marine waters. From spring to fall, they can be seen in the Puget Sound and Georgia Basin.

Food: Fish, sharks, birds, and marine mammals such as seals, sea lions, and even whales and other dolphins

Den: No den; young are birthed live in open water.

Young: Usually, a single calf is born after a 17-month pregnancy. Calves nurse for a year and will remain close to mom for the first 2 years of life. They reach reproductive maturity at around 10 years.

Predators: Other killer whales and humans

Tracks: Fully aquatic marine mammal that leaves no tracks

Orcas are overall black with white patches around the eyes, mouth, and underside. They have a gray or white patch behind the dorsal fin called a saddle patch. Males are larger than females and have larger pectoral fins, dorsal fins, and tail flutes.

This animal is fully aquatic, so it does not leave tracks.

Did you know?

Pikas are the smallest member of the Rabbit family. Pikas are also called "whistling hares" because they give out loud warning calls in the presence of danger; in fact, many times you will hear a pika before you see it.

Size Comparison

Most Active

Track Size

Pika

Ochotona princeps

Size: 6½–8½ inches long; weighs 4–6½ ounces

Habitat: Mountains above the tree line, rock faces, cliffs, alpine terrain, slopes of mountain meadows, and hillsides

Range: The pika lives in southwest Canada and the western US. In Washington, it can be found throughout the Cascade Range and the northeast portion of the state.

Food: Pikas are herbivores that eat grasses, weeds, and wildflowers. They will sometimes eat their own waste (poop) that is rich in protein and energy.

Den: Dens are formed in rock piles and are no larger than 3 feet deep.

Young: After 30 days of pregnancy, 2 litters of kits (ranging from 2–6 young) are born each year. The young depend on their mother at birth. Young are weaned around a month after birth and can breed the following year.

Predators: Prairie falcons, red foxes, golden eagles, coyotes, American martens, red-tailed hawks, bobcats, and long-tailed weasels

Tracks: Front feet are ¾ inch long and ⅗ inch wide with five toes. Hind feet are 1–1¼ inches long and 2½–3½ inches wide with four toes.

Pikas are small mammals with an oval-shaped, stout body. They have big ears and short legs. Pikas have thick fir that is brown to tannish with an overlaying of black on their back.

Did you know?

Elk are known to be the loudest of all cervids (Deer family). Males produce a low-pitched bellow or roar, called a bugle. Bugling is a technique that involves both roaring and whistling at the same time. Elk use their bugle or bugling to attract mates or announce territories during the fall mating season. Their bugles can be heard over long distances.

Size Comparison

Most Active

Track Size

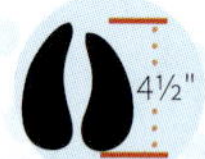

Roosevelt/Rocky Mountain Elk

Cervus candensis roosevelt/Cevrus candadensis nelsoni

Size: 5–8 feet tall; weighs 377–1,095 pounds

Habitat: Open woodlands, mountain areas, shrublands, coniferous swamps, and hardwood forests

Range: Found throughout the western US, portions of the Southeast, and in Canada. In Washington, the Roosevelt elk are found in the western areas that border the Cascade Range. Rocky Mountain elk are found mainly in the eastern portion of the state.

Food: Elk are herbivores that eat grasses; flowers; and leaves from trees like cedar, red maple, and basswood.

Den: No den; mother elk will hide young calves in tall grasses.

Young: Calves are born after 240–265 days. At birth, calves weigh around 30 pounds and have spots through the first summer. Separation from mother's milk happens around the 60-day mark, but calves will continue to get care and protection from mom for around a year.

Predators: Mountain lions, Mexican gray wolves, and bears. Calves may fall victim to bobcats and coyotes.

Tracks: Front tracks of an adult are about 4¾ inches long and wide. Hind foot tracks are 4½ inches long and 3½ inches wide. Two toes are on each foot.

In the summer and spring, they are lighter brown to tan; in the winter, elk are a deep dark brown. During both seasons, they have a cream or off-white rump. They sport a darker tone on the head, neck, belly, and legs. Rocky Mountain elk are paler and skinnier than Roosevelt elk.

Did you know?

Snowshoe hares can run over 30 miles per hour and can jump over 12 feet in a single leap. Their name comes from their hind feet, which are large and furry and look (and act) like snowshoes.

Size Comparison

Most Active

Track Size

Snowshoe Hare

Lepus americanus

Size: 17–22 inches long; tail is 2 inches long; weighs 3–4½ pounds

Habitat: Woody areas, swamps, open fields, and forest bogs

Range: They are found widespread in the state of Washington, the northern US into the Appalachian Mountains, and on the Pacific Coast.

Food: Grasses, flowers, bark, twigs, evergreen needles; they sometimes will eat the remains of other snowshoe hares.

Den: None

Young: They give birth to 2–4 young (known as leverets); during the day, young will hide in different places and will only come together with the mother to nurse (drink milk) for a few short minutes. Young are mature one year after birth.

Predators: Bobcats, coyotes, mink, foxes, and owls and other birds of prey

Tracks: Their hind feet are larger than the front.

The snowshoe hare is a medium-size, rabbit-like animal with brown fur in the summer and a white coat in the winter. They are mostly active during the dawn and dusk hours of the day (crepuscular).

Did you know?

Skunks help farmers! They save farmers money by feeding on rodents and insects that destroy crops. When skunks spray, they can aim really well! When threatened, a skunk will aim its tail towards the threat and spray a stinky musk into the target's face or eyes.

Size Comparison

Most Active

Track Size

Hibernates

Striped Skunk

Mephitis mephitis

Size: 17–30 inches long; weighs 6–13 pounds

Habitat: Woodlands, prairies, and suburban areas

Range: Common throughout the state of Washington; they can be found throughout the US and into Canada and the northern parts of Mexico.

Food: Omnivores (eaters of meat and plants), they eat eggs, fruits, nuts, small mammals, carrion (dead things), insects, amphibians, small reptiles, and even garbage.

Den: Skunks prefer short and shallow natural dens, or dens abandoned by other animals, but will dig dens 3–6 feet long and up to 3 feet deep underground. Dens have multiple hidden entrances, and rooms are usually lined with vegetation.

Young: They have 4–5 young (kits) that are blind at birth; at around 3 weeks they gain vision and the ability to spray.

Predators: Raptors and large carnivores

Tracks: Their front feet have five long, curved claws used for digging; the hind foot also has five toes and is longer and skinnier than the front foot.

The striped skunk is a cat-size, nocturnal (active at night) mammal with black fur and two white stripes that run the entire length of the body. The stripe pattern is usually distinctive to each skunk.

Did you know?

American robins have a great sense of hearing. They hunt for earthworms underground using only their hearing. Robins are opportunistic feeders in urban (city) areas; they will wait for lawns to be disturbed by mowers, sprinklers, or rain, and then feed on the worms that have emerged.

Nest Type

Most Active

American Robin

Turdus migratorius

Size: 9–11 inches long; wingspan of 17 inches; weighs 2½–3 ounces

Habitat: Cities, forests, and lawns

Range: They are widespread throughout Washington as year-round residents. They can be found throughout North America, except for the extreme north of Canada.

Food: Fruits, earthworms, beetle grubs, caterpillars, insects, and grasshoppers

Nesting: April to July

Nest: Cup-shaped nests are exclusively built by the female 5–14 feet off the ground in bushes or trees. Nests are constructed of grass, paper, twigs, and feathers. A new nest is built for each set of eggs.

Eggs: 3–5 sky-blue eggs

Young: Eggs hatch after 14 days of incubation; chicks hatch blind and mostly without feathers. Hatchlings (chicks) leave the nest after 2 weeks but will continue to beg for food from parents.

Predators: Snakes, crows, cats, foxes, raccoons, squirrels, raptors, and weasels

Migration: Year-round resident

American robin males have a dark black-to-gray head with a yellow bill, a brown back, a rusty-orange chest, and a whitish ring around the eyes. Females are similar in color but are not as bright as males, and they usually have a brownish head.

Did you know?

Anna's hummingbirds are only found in North America. Their heart beats around 1,260 beats per minute, and they can shake their bodies over 50 times per second while flying, in order to quickly dry or remove dust from their feathers. Anna's hummingbirds are the only hummingbirds with a red or pink crown.

Nest Type

Most Active

Anna's Hummingbird

Calypte anna

Size: 4–4¼ inches long; wingspan of 4½–4¾ inches; weighs ⅒–⅕ ounce

Habitat: Shrublands, farmlands, woodlands, forests, urban areas, mountains, and open wooded areas

Range: They can be found from southern Canada to northern Baja California, Mexico, and in the western US. They are found throughout the western and central areas of Washington.

Food: They are herbivores that sip nectar from flowers; they will also sometimes eat insects and spiders.

Nesting: Mid-December to June

Nest: They make a small cup nest (1 inch tall and 1½ inches wide) out of feathers, plant materials, and fur. Nests are held together with spiderwebs.

Eggs: 2 small white eggs are laid per brood.

Young: Chicks hatch about 2–2½ weeks after eggs are laid with a little down and eyes closed. They fledge from the nest at 18–24 days and reach maturity at a year old.

Predators: Snakes, birds of prey, roadrunners, and pet cats

Migration: They do not migrate.

Anna's hummingbirds are small, brightly colored birds. Their back is green with an iridescent bronze–golden-brown sheen. They have a gray breast and stomach and green sides. The males have iridescent ruby-pink-to-deep-magenta feathers on their head and throat. They have skinny, long, straight bills. Females and juveniles have a dull-green head, mostly gray throat, gray breast and stomach, and a rounded tail. Males have a slightly forked tailed.

Did you know?

The bald eagle is an endangered species success story! The bald eagle was once endangered due to a pesticide called DDT that weakened eggshells and caused them to crack early. Through the banning of DDT and other conservation efforts, the bald eagle population recovered, and it was removed from the Endangered Species List in July of 2007.

Nest Type

Most Active

Bald Eagle

Haliaeetus leucocephalus

Size: 3½ feet long; wingspan of 6½–8 feet; weighs 8–14 pounds

Habitat: Forests and tree stands (small forests) near river edges, lakes, seashores, and wetlands

Range: They are year-round residents in Washington; they are found throughout much of the US.

Food: Fish, waterfowl (ducks), rabbits, squirrels, muskrats, and deer carcasses; will steal food from other eagles or osprey

Nesting: Eagles have lifelong partners that begin nesting in fall, laying eggs between November–February.

Nest: They build a large nest out of sticks, high up in trees; the nest can be over 5 feet wide and over 6 feet tall, often shaped like an upside-down cone.

Eggs: 1–3 white eggs

Young: Young (chicks) will hatch at around 35 days; young will leave the nest at around 12 weeks. It takes up to 5 years for eagles to get that iconic look!

Predators: Few; collisions with cars sometimes occur.

Migration: Year-round residents in Washington

Adult bald eagles have a dark-brown body, a white head and tail, and a golden-yellow beak. Juvenile eagles are mostly brown at first, but their color pattern changes over their first few years. A bald eagle can use its wings as oars to propel itself across bodies of water.

Did you know?
Black-capped chickadees have a unique strategy for surviving winter. The area of the brain that aids in memory (the hippocampus) temporarily gets bigger in preparation for winter. This allows them to remember where they hid or cached seeds.

Nest Type

Most Active

Black-capped Chickadee

Poecile atricapillus

Size: 5½–7½ inches long; wingspan of 8 inches; weighs about half an ounce

Habitat: Forests, woodland edges, and suburban and urban areas

Range: They are year-round residents of Washington and can be found in the northern United States and Canada.

Food: Caterpillars, insects, seeds, spiders, and berries

Nesting: April to August

Nest: Chickadees utilize old woodpecker holes or make their own cup-shaped nests in tree cavities that have been weakened by rot.

Eggs: 4–6 eggs that are white with brown spots

Young: Eggs hatch 12–13 days after they are laid; chicks leave the nest around 15 days after hatching; chickadee parents continue feeding the young for another 5–6 weeks.

Predators: Hawks, owls, shrikes, raccoons, house cats left outside, and other mammals

Migration: They do not migrate.

A black-capped chickadee has a gray body with a black cap, or top of head, and a black throat and beak; they have white cheeks and light bellies.

Did you know?

Buffleheads are the smallest diving ducks in North America. Their range is limited to areas where northern flickers live because buffleheads nest in holes excavated by northern flickers (and sometimes pileated woodpeckers).

Nest Type

Most Active

Migrates

Bufflehead

Bucephala albeola

Size: 12½–15¾ inches long; wingspan of 21½ inches; weighs 9½–22½ ounces

Habitat: Coastal areas, lakes, beaches, coves, marshes, woodlands, and forests

Range: They can be found throughout much of the US and Canada. In Washington, they are found as winter residents throughout much of the state and as a breeding resident in the central portion of the state.

Food: They are carnivores that eat crab, shrimp, snails, clams, and aquatic insects and invertebrates.

Nesting: Late winter to April

Nest: Utilizes nesting cavities excavated or dug out by northern flickers and sometimes pileated woodpeckers. Nests are lined with down feathers.

Eggs: On average, 8 cream-colored eggs are laid per clutch.

Young: Chicks hatch 28–30 days or so after laying. Hatchlings fledge after 45–50 days and will reach reproductive maturity at around 2 years old.

Predators: Black bears, American mink, bald eagles, peregrine falcons, owls, weasels, and hawks

Migration: Some migrate to breed and return to Washington to overwinter.

Buffleheads are small, big-headed ducks. The males are larger and more colorful than the females. Males' heads are iridescent colors of greens and purples when viewed up close. From afar, the head appears to be black and white. Breeding males have a bluish bill. Both sexes have a wide, short bill. Females and non-breeding males are grayish brown with a white cheek patch.

Did you know?

Canada geese sometimes travel over 600 miles in a day. They fly in a V formation, which allows them to travel long distances without stopping because they can switch positions. As the lead bird gets tired, it drops to the back of the line and a new bird leads. The V formation helps them communicate and helps prevent collisions.

Nest Type

Most Active

Migrates

Canada Goose

Branta canadensis

Size: 2–3½ feet long; wingspan of 5–6 feet; weighs 6½–20 pounds

Habitat: Ponds, marshes, lakes, parks, and farm fields

Range: In Washington, they are year-round residents in much of the state, while in other parts, they are breeding residents; they are widespread in the rest of the US.

Food: Omnivores, they eat grasses, aquatic insects, seeds, and some crops, like corn or alfalfa.

Nesting: March to April

Nest: Nests are made on the ground, on elevated areas near the water, or sometimes on a muskrat mound. Nest sites are picked with protection in mind; areas that have clear views and vantage points are more likely to be used.

Eggs: 2–8 cream-colored eggs that are 3 inches long and about 2½ inches wide

Young: Goslings hatch about a month after being laid. They are born with yellow down feathers that they lose as they get older. At the time of hatching, they can swim and walk.

Predators: Mink, raccoons, foxes, dogs, and great horned owls

Migration: Nonmigratory bird in much of Washington; in other areas of the state, they migrate to breed.

The Canada goose is recognizable by its famous honk and body pattern of brown feathers with a black neck, head, bill, and even feet. They have white cheek feathers.

Common Raven

American Crow

Did you know?

Ravens can mimic other bird species and even dogs and humans in cities. They will work together to steal eggs or chicks out of other birds' nests. Crows in the Pacific Northwest are smaller than those in the eastern US and have a deeper voice. Crows are smart; they have been observed using a cup to get water and wetting dry food. They will recognize individual humans who have caused them harm.

Nest Type

Most Active

Common Raven (CR)/ American Crow (AC)

Corvus corax/Corvus brachyrhynchos

Size: CR: 22–27¼ inches long; wingspan of 45¾–46½ inches; weighs 24¼–57¼ ounces. AC: 15¾–21 inches long; wingspan of 33½–39½ inches; weighs 11¼–22 ounces

Habitat: CR: Forests, beaches, islands, sagebrush, mountains, deserts, grasslands, and farm fields. AC: Parking lots, fields, lawns, forests, dumps, and woodlands

Range: CR: Across much of North America, except in the Midwest and Southeast. AC: Across the US as year-round and nonbreeding residents, as well as Canada as breeding residents. Both species are year-round residents throughout Washington.

Food: Carrion, rodents, bird eggs, insects, nuts, berries, fish, and sometimes even pet food

Nesting: AC: February–May; CR: February

Nest: CR: Cup nest on cliff edge, tree, or human-made structure. AC: Cup nest is made of twigs.

Eggs: CR: 3–7 green, blue, or olive eggs, often with marks. AC: 3–9 pale-to-olive-green eggs with blotches

Young: Young hatch after 18 days with little tufts of feathers and their eyes closed. They will fledge within 20–40 days.

Predators: Pet cats, owls, hawks, coyotes, and humans

Migration: They do not migrate.

Ravens and crows sport completely black feathers. Their feet, legs, and bill are also black. Ravens are slightly larger than crows and have bigger beaks and a wedge-shaped tail.

Did you know?

The dark-eyed junco is one of the most common birds in North America. They can be found from coast to coast. At one time, the dark-eyed junco was thought to be five different species, but now it's considered one species with several subspecies.

Nest Type

Most Active

Migrates

Dark-eyed Junco

Junco hyemalis

Size: 5½–6½ inches long; wingspan of 7–10 inches; weighs about an ounce

Habitat: Woodlands, shrub areas, and forest areas

Range: They can be found throughout the United States and Canada. In Washington, they are year-round residents in much of the state; in other parts, they are nonbreeding residents.

Food: Seeds, insects, spiders, and berries

Nesting: Starts in April

Nest: A cup-shaped nest is made of grass, animal hair, moss, or bark; it's usually built in a depression (a shallow hole) on the ground and hidden by vegetation and other natural materials like rocks or logs.

Eggs: 4 eggs that are bluish or gray with a slight gloss and brown spots

Young: The eggs are incubated by the female for 12–13 days. The chicks leave the nest between 11 and 14 days after hatching and become reproductively mature at 1 year of age.

Predators: Chipmunks, shrikes, owls, cats, weasels, squirrels, and hawks

Migration: Year-round residents that do not migrate; winter population will migrate north to breeding areas.

The dark-eyed junco is a smoky-gray bird with a white belly and pink bill.

Pelagic

Double-crested

Did you know?

Cormorants do not have waterproof feathers like other aquatic birds, which is why you may see one on a rock or post with its wings spread: it's drying itself off after a successful hunt. Cormorants' bills curve at the end. Pelagic cormorants are smaller than double-crested cormorants and are rarely found inland. They can dive over 120 feet underwater to catch fish.

Nest Type

Most Active

Migrates

Double-crested (DC)/ Pelagic Cormorant (PC)

Nannopterum auritum/Phalacrocorax pelagicus

Size: DC: 26–35 inches long; wingspan of 45–48½ inches; weighs 5–5½ pounds. PC: 20–30 inches long; wingspan of 39–47 inches; weighs 3–5¼ pounds

Habitat: DC: Lakes, rivers, swamps, and coastal waters. PC: Rocky areas on the coast and islands

Range: DC: Across North America and Washington. PC: Along the western portions of North America. In Washington, they are found along the coast.

Food: Fish, insects, snails, and crayfish

Nesting: DC: April–August. PC: May–July

Nest: DC: On the ground or in a tree. PC: Shallow bowl

Eggs: DC: 4 light-bluish-white eggs. PC: 1–8 bluish- or white-hued green eggs

Young: DC: Young usually hatch in 25–28 days. They can reproduce at around 2 years. PC: Chicks hatch about a month after laying. They fledge in about 50 days.

Predators: Coyotes, foxes, raccoons, eagles, and owls

Migration: DC: year-round residents except for migrators in the central part of the state and breeding visitors in the eastern corner and western tip
PC: year-round residents along the coast

Adult double-crested cormorants have topaz-colored eyes. During breeding season, adults may have a "double crest" of black-and-white feathers and are blackish with a green or purple gloss. Breeding pelagic cormorants have white patches behind the wings. They are overall black with a green sheen on the body, purple sheen on the head, and a red patch on the chin.

Did you know?

Golden eagles are the largest birds of prey that actively hunt in North America! They are North America's second-largest bird of prey after the California condor, which is a scavenger. Golden eagles also have the widest range of all eagles. They can be found across North America, Africa, Europe, and Asia. Several countries have named the golden eagle as their national bird, including Mexico, Albania, Kazakhstan, and Austria.

Nest Type

Most Active

Golden Eagle

Aquila chrysaetos

Size: 26–33 inches long; wingspan of 6–8 feet; weighs 7½–9 pounds

Habitat: Shrublands, forests, mountains, cliffs, canyons, grasslands, rocky areas, and woodlands

Range: Widespread across North America throughout various seasons. They are found throughout Washington.

Food: Birds, goats, sheep, coyotes, pronghorns, badgers, reptiles, deer, fish, bobcats, and small mammals

Nesting: January–May

Nest: Nests on cliffs, trees, and buildings. Nests are made of sticks and other plant materials; they will sometimes include animal bones and human objects.

Eggs: 2–3 cream-to-light-pink eggs are laid.

Young: Eaglets hatch about 40–45 days after eggs are laid. They learn to fly at around 10 weeks. They will reproduce at around 4–7 years.

Predators: Wolverines and grizzly bears are the major predators of chicks. Humans are often the cause of death due to habitat loss and hunting practices.

Migration: Year-round residents in central Washington; overwintering residents will migrate to breeding grounds.

Golden eagles are covered in dark-brown feathers. They have a golden neck and sides of their face, which is where they get their name. They have broad wings that stretch over 7 feet across. The tail is faded brown to dark brown. They have deep-brown eyes, a black-tipped bill, and black claws. Their feet are yellow, and their legs are covered with feathers. Immature eagles have patches of white-to-buffy-colored feathers.

Did you know?

The great blue heron is the largest and most common heron species. A heron's eye color changes as it ages. The eyes start out gray but transition to yellow over time. Great blue herons swallow their prey whole.

Nest Type

Most Active

Migrates

Great Blue Heron

Ardea herodias

Size: 3–4½ feet long; wingspan of 6–7 feet; weighs 5–7 pounds

Habitat: Lakes, ponds, rivers, marshes, lagoons, wetlands

Range: They can be found throughout the United States and down into Mexico. In Washington, they are year-round residents throughout much of the state, and the rest are visitors during migration.

Food: Fish, rats, crabs, shrimp, grasshoppers, crayfish, other birds, small mammals, snakes, and lizards

Nesting: May–August

Nest: 2–3 feet across and saucer shaped; often grouped in large rookeries (colonies) in tall trees along the water's edge. Nests are built out of sticks and are often located in dead trees more than 100 feet above the ground; nests are used year after year.

Eggs: 3–7 pale bluish eggs

Young: Chicks will hatch after 28 days of incubation; young will stay in the nest for around 10 weeks. They reach reproductive maturity at just under 2 years.

Predators: Eagles, crows, gulls, raccoons, bears, and hawks

Migration: Some herons will migrate to breeding grounds, while the rest are year-round residents.

The great blue heron is a large wading bird with blue and gray upper body feathers; the belly area is white. They have long yellow legs that they use to stalk prey in the water. Great blue herons are famous for stalking prey at the water's edge; their specially adapted feet keep them from sinking into the mud!

Did you know?

They are the tallest owls in North America and the world's largest species of owl by length. They have been known to successfully defend their nests from large predators like black bears! Great gray owls have excellent hearing and can hunt prey buried under 2 feet of snow just by sound.

Nest Type

Most Active

Great Gray Owl

Strix nebulosa

Size: 24–33 inches long; wingspan of 55–59¾ inches; weighs 20½–60 ounces

Habitat: Grasslands, shrublands, woodlands, forests, wetlands, and meadows

Range: They breed in North America from as far east as Quebec to the Pacific Coast and Alaska. They are year-round residents from central Washington eastward.

Food: They are carnivores that mainly eat small mammals like shrews and voles, but they also eat ducks, hares, and quails.

Nesting: March to May

Nest: They use old nests previously built by large birds. They will also nest in the tops of broken trees and in cavities in large trees.

Eggs: 4 dull-white eggs are laid per clutch.

Young: Owlets hatch blind and helpless after a month of incubation. They begin to leave the nest in 2–3 weeks by jumping to nearby branches. They reach maturity around 2–3 years old.

Predators: They have few natural predators due to their size.

Migration: Does not migrate

Great gray owls have large, round heads with yellow eyes and a yellow bill. Their face is cup- or disk-shaped with no ear tufts. They have a white X pattern on the middle of their face. They have a long tail. The back of the owl has more gray and brown with less white than the front.

Did you know?

A great horned owl can exert a crushing force of over 300 pounds with its talons. Despite its name, the great horned owl doesn't have horns at all. Instead, the obvious tufts on its head are made of feathers. Scientists aren't sure exactly how the tufts function, but they may help them stay hidden.

Nest Type

Most Active

Great Horned Owl

Bubo virginianus

Size: Up to 23 inches long; wingspan of 45 inches; weighs 3 pounds

Habitat: Woods; swamps; desert edges; as well as heavily populated areas such as cities, suburbs, and parks

Range: They are found throughout the continent of North America. They are year-round residents in Washington.

Food: They eat a variety of foods, but mostly mammals. Sometimes they eat other birds as well.

Nesting: They have lifelong partnerships, with nesting season starting in early winter; egg-laying starts in mid-January to February.

Nest: Nests are found 20–50 feet off the ground. They tend to reuse nests from other raptors or hollowed-out trees.

Eggs: The female lays 2–4 whitish eggs. Eggs are incubated for around 30 days.

Young: Young can fly at around 9 weeks old. The parents care for and feed young for several months.

Predators: Young owls are preyed upon by foxes, coyotes, bears, and opossums. As adults, they are rarely attacked by other birds of prey, such as golden eagles and goshawks.

Migration: Great horned owls are not regular migrators, but some individuals will travel south during the winter.

They are bulky birds with large ear tufts, a rusty brown-to-grayish face with a black border, and large bright eyes. The body color tends to be brown; the wing pattern is checkered with an intermingled dark brown. The chest and belly areas are light brown and have white bars.

Did you know?

Hairy woodpeckers can hear insects traveling under the tree bark. Downy woodpeckers are the smallest woodpecker species in North America. Downy woodpeckers have a built-in mask, or special feathers, near their nostrils that helps them to avoid breathing in wood chips while pecking.

Nest Type

Most Active

Hairy/Downy Woodpecker

Leuconotopicus villosus/Dryobates pubescens

Size: Hairy: 7–10 inches long; wingspan of 13–16 inches; weighs 3 ounces. Downy: 5½–7 inches long; wingspan of 10–12 inches; weighs less than an ounce

Habitat: Forested areas, parks, woodlands, and orchards

Range: Hairy: Year-round residents in Washington and a small corner of the western tip of Texas that borders New Mexico. Downy: Found throughout much of the state

Food: Hairy: Beetles, ants, caterpillars, fruits, and seeds. Downy: Beetles, ants, galls, wasps, seeds, and berries

Nesting: Hairy: March to June. Downy: January to March

Nest: In both woodpecker species, pairs will work together to create a cavity. Both parents also help to incubate eggs.

Eggs: Hairy: 3–7 white eggs. Downy: 3–8 white eggs

Young: Hairy woodpeckers' eggs will hatch 2 weeks after being laid and then fledge (develop enough feathers to fly) after another month. Downy woodpeckers' eggs will hatch after about 12 days and fledge 18–21 days after hatching. Both species hatch blind and featherless.

Predators: American kestrels, snakes, sharp-shinned hawks, pet cats, rats, squirrels, and Cooper's hawks

Migration: Woodpeckers are mostly year-round residents.

Hairy woodpeckers and downy woodpeckers look strikingly similar with their color pattern. One way to distinguish them is to look at the size of the body and bill. The downy woodpecker is smaller than the hairy woodpecker and has a shorter bill. If you look at the tail feathers of the two species, you will also see that the hairy woodpecker does not have black spots, while the downy's tail does.

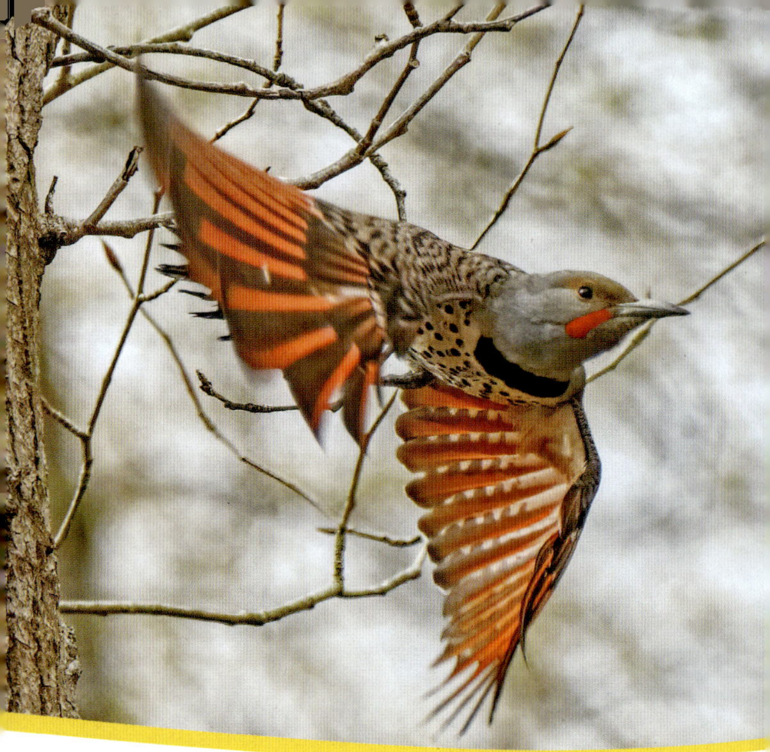

Did you know?

The northern flicker has the longest tongue of all birds in North America. Their tongues can extend as far as 2 inches past the tip of their bill. Northern flickers use a technique called anting where they will let ants crawl on them to help keep themselves clean. The ants release a chemical called formic acid that gets rid of parasites that are on the bird's feathers.

Nest Type

Most Active

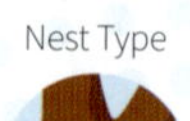

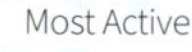

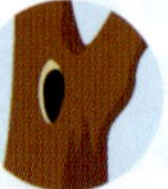

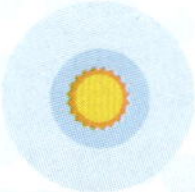

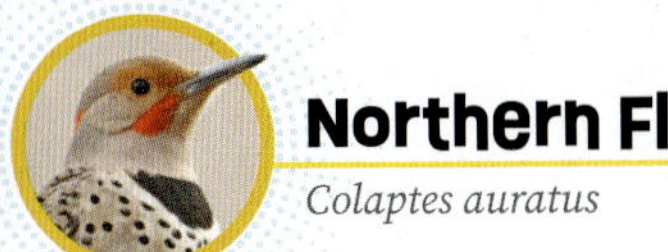

Northern Flicker

Colaptes auratus

Size: 11–12¼ inches long; wingspan of 16½–20 inches; weighs 4–5½ ounces

Habitat: Woodlands, yards, marsh and swamp edges, and parks

Range: They can be found across North America. In Washington, they are year-round residents throughout the state.

Food: Omnivores, they eat insects, berries, seeds, and snails.

Nesting: February to July

Nest: They excavate nest holes in the trunks or branches of dead or diseased trees.

Eggs: 5–8 all-white eggs

Young: Chicks hatch naked with eyes closed after an 11-day incubation by both parents. Around the 17-day mark, they start hanging on the walls of the cavity. They will fledge within a week but will remain close to parents for around 25–30 days.

Predators: Snakes, birds of prey, squirrels, pet cats, and raccoons

Migration: They do not migrate.

The northern flicker has a slender body, round head, and long bill that is partially curved. They are mostly brown with black spots and bars across their body. They have a white rump patch. Flickers of the western portion of North America are known as red shafted due to the red undersides of their wing and tail feathers. Flickers in the east are known as yellow shafted. Red shafted also have a gray face and brown crown, and they do not have a crescent on the back of their neck like the yellow shafted.

Did you know?

The osprey is nicknamed the "fish hawk" because it is the only hawk in North America that mainly eats live fish. An osprey will rotate its catch to put it in line with its body, pointing headfirst, which allows for less resistance in flight as the air travels over the fish.

Nest Type

Most Active

Migrates

Osprey

Pandion haliaetus

Size: 21–23 inches long; wingspan of 59–71 inches; weighs 3–4½ pounds

Habitat: Near lakes, ponds, rivers, swamps, and reservoirs

Range: They are found throughout the US. In Washington, they are breeding residents and migration visitors.

Food: Feeds mostly on fish; they sometimes eat mammals, birds, and reptiles if there are few fish.

Nesting: For ospreys that migrate, egg-laying happens in April and May. The female will take on most of the incubation of the eggs, as well as the jobs of keeping the offspring warm and providing protection.

Nest: Platform nests are made out of twigs and sticks on trees, snags, or human-made objects.

Eggs: The mother lays 1–3 cream-colored eggs with splotches of browns and pinkish reds.

Young: Chicks hatch after around 36 days and have brown-and-white down feathers. Ospreys fledge around 50–55 days after hatching.

Predators: Owls, eagles, foxes, skunks, raccoons, and snakes

Migration: Ospreys migrate to overwintering grounds in the southern US or Central and South America in the fall, and to breeding grounds in the spring.

Ospreys are raptors, and they have a brown upper body and white lower body. The wings are brown on the outside and white on the underside, with brown spotting and streaks toward the edge. The head is white with a brown band that goes through the eye area, highlighting the yellow eyes.

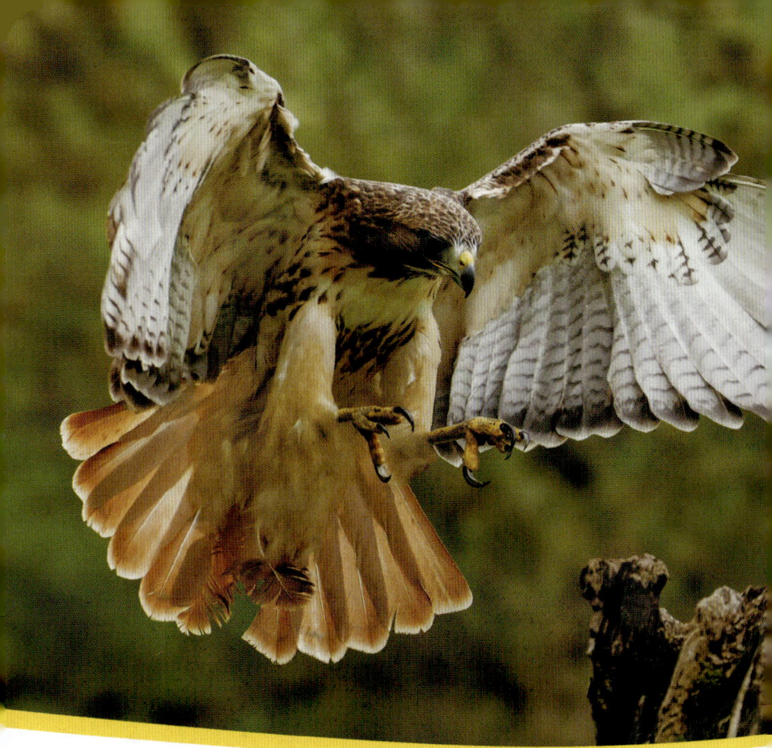

Did you know?

The red-tailed hawk is the most abundant hawk in North America. The red-tailed hawk's scream is the sound effect that you hear when soaring eagles are shown in movies. Eagles do not screech like hawks, so filmmakers use hawk calls instead! Red-tailed hawks can't move their eyes, so they have to move their entire head in order to get a better view around them.

Nest Type

Most Active

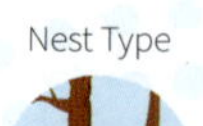

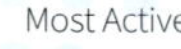

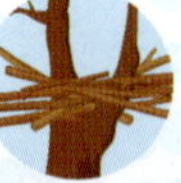

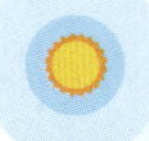

Red-tailed Hawk

Buteo jamaicensis

Size: 19–25 inches long; wingspan of 47–57 inches; weighs 2½–4 pounds

Habitat: Deserts, woodlands, grasslands, and farm fields

Range: Throughout North America; in California, they are widespread throughout the state as year-round residents.

Food: Rodents, birds, reptiles, amphibians, bats, and insects

Nesting: Hawks mate for life; nesting starts in March.

Nest: Both the male and female help build a large cup-shaped nest, which can be over 6 feet high and 3 feet across; the nest is made of sticks and branches. Nests are built at forest edges mostly in the crowns of trees, but hawks will also nest on windowsills and other human-made structures.

Eggs: 1–5 eggs; the insides of eggs are a greenish color.

Young: They start to fly after 5–6 weeks, and it takes around 10 weeks for the hatchlings to leave the nest.

Predators: Owls and crows

Migration: Does not migrate

Red-tailed hawks are named for their rusty-red tails! They have brown heads and a chest that's cream to light brown with brown streaking in the form of a band. Red-tailed hawks are highly territorial, and throughout the day they will take to the air to look for invaders.

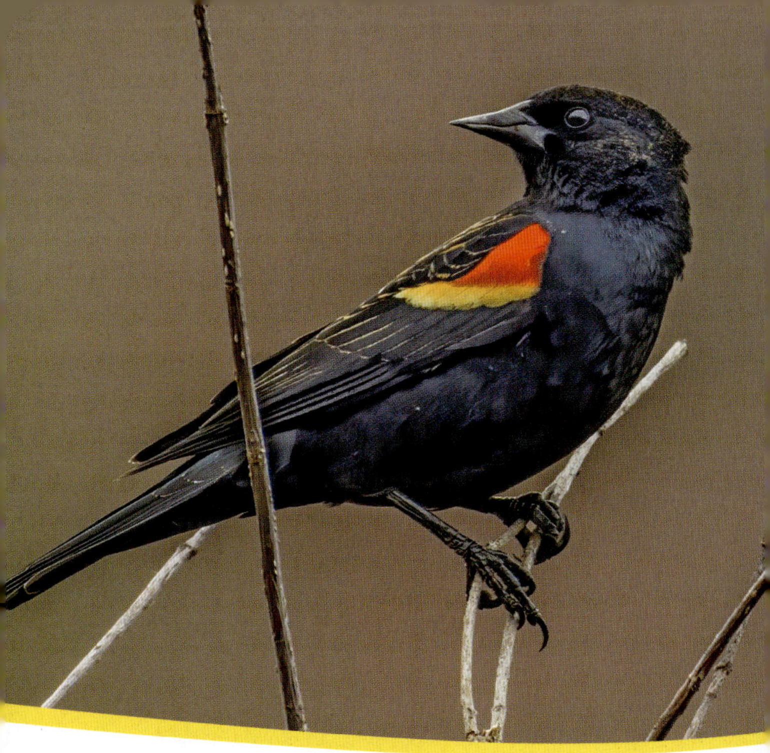

Did you know?

Red-winged blackbirds are one of the most abundant songbirds in the United States. Sometimes their winter roost (colony) can have several thousand to up to a million birds. In many areas, red-winged blackbirds are considered a pest because of their love of grain and seeds from farm fields. In others, they are welcomed because they eat insects that are considered pests to farmers.

Nest Type

Most Active

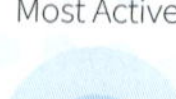

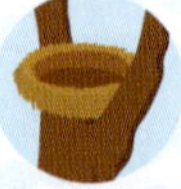

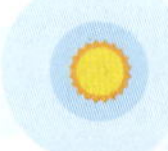

Red-Winged Blackbird

Agelaius phoeniceus

Size: 7–9½ inches long; wingspan of 13 inches; weighs 2 ounces

Habitat: Marshes, lakeshores, meadows, parks, and open fields

Range: In Washington, found as year-round residents; ranges from central Canada through the US and into Mexico

Food: Dragonflies, spiders, beetles, snails, seeds, and fruits

Nesting: February to August

Nest: Female builds a cup from plant material.

Eggs: 3–4 eggs that come in a variety of colors, from pale blue to gray with black-and-brown spots or streaks

Young: Chicks hatch blind and naked after around 12 days of incubation. Hatchlings will leave the nest after 12 days but will continue to receive care for another 5 weeks.

Predators: Raccoons, mink, marsh wrens, and raptors

Migration: Washington populations do not migrate.

Red-winged blackbird males are a sleek black with an orangish-red spot that overlays a dandelion-yellow spot on the wings. Females have a combination of dark-brown and light-brown streaks throughout the body. Male red-winged blackbirds spend much of breeding season defending their territory from other males and attacking predators or anything else that gets too close to the nest.

Did you know?

Steller's jays are the only birds west of the Rockies that have a crest. Steller's jays can (and will) mock or mimic sounds of their surroundings. Often, they will imitate the calls of birds of prey to scare off other birds from food sources.

Nest Type

Most Active

Steller's Jay

Cyanocitta stelleri

Size: 11¾–13½ inches long; wingspan of 17¼ inches; weighs 3½–5 ounces

Habitat: Woodlands, forests, mountainous areas, and parks

Range: They can be found from southern Alaska through the western US into Mexico and Central America. In Washington, they are found throughout much of the state as year-round residents.

Food: They are omnivores that feed on berries, seeds, nuts, and other plant material, as well as insects, eggs, small rodents, lizards, snakes, and even other bird nestlings.

Nesting: Late March to early July; peak in April to May

Nest: The cup nest is made of stems, moss, sticks, and leaves and is lined with fur, roots, and pine needles.

Eggs: 2–6 bluish-green eggs with brown, purple, or olive spotting are laid per clutch.

Young: Chicks hatch 16 days after laying, featherless and with their eyes closed. They learn to fly in 18 days.

Predators: Snakes, birds of prey, raccoons, crows, and pet cats

Migration: They do not migrate.

Steller's jays are large, stout songbirds. They have a big head, rounded wings, and a long tail. They have a triangular crest on the top of their head and a long bill. Their head is black on top, transitioning to blue on the lower breast and belly. They have blue wings and a tail with dark barring. Their color can vary based on location; birds in eastern California have white markings on the head, birds closer to the west have light-blue markings, and birds of the Pacific Coast have light or no markings at all.

Did you know?

Tufted puffins are amazing flyers; they can reach speeds of 40 miles per hour and can beat their wings at over 300 beats per minute. Their impressive movements are not limited to the air. Because of their dense bones, they can dive up to 200 feet at one time and catch up to 20 small fish in their bill. On land, they use their feet and bill to dig nesting burrows that can reach over 5 feet deep.

Nest Type

Most Active

Migrates

Tufted Puffin

Fratercula cirrhata

Size: 14¼–15¾ inches long; wingspan of 30 inches; weighs 18¼–35¼ ounces

Habitat: Open ocean, offshore islands, cliffs, and coastal areas

Range: They are widespread in the north Pacific Ocean and nest on coastlines and offshore islands from Alaska to southern California. They can be found along the coast of Washington during the spring and summer.

Food: Small fish, squid, octopus, crabs, zooplankton, and jellyfish

Nesting: Mid-April to early September

Nest: Nests are made in burrows on grassy cliffs.

Eggs: A single white egg is laid per year.

Young: After 45 days, the chick hatches, covered in down feathers, with the ability to walk. Both parents care for the young for 6–7 weeks after hatching. When the chick is ready to leave the nest, it will do so at night and walk and flap its way to the sea.

Predators: Orcas, bald eagles, sea lions, foxes, owls, and rats; gulls will steal food from puffins.

Migration: Migrates from the open ocean to coastal areas to breed

Tufted puffins are the largest species of puffins. During the breeding season, tufted puffins have an orange-and-yellow bill, a white-masked face, yellow eyes, and backward-facing yellow feather tufts behind each eye. They have black-to-dark-brown bodies, short wings, and orange feet. Overwintering adults have some orange on their bills, but they lose their white mask and tufts. Juveniles are brownish gray on top and paler below with a smaller yellow bill.

Did you know?

Varied thrushes have a song that is a single note long. The varied thrush forages by moving leaves as it hops around on the ground, flying to a viewing spot, then flying down to a new area.

Nest Type

Most Active

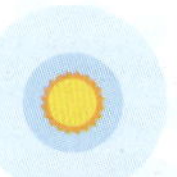

Migrates

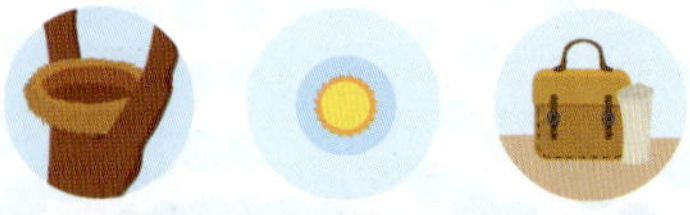

Varied Thrush

Ixoreus naevius

Size: 7½–10¼ inches long; wingspan of 13½–15 inches; weighs 2¼–3½ ounces

Habitat: Forests, parks, gardens, ravines, and backyards

Range: They are common in the Cascade Range as well as the Pacific Coast. They are nonbreeding residents in central Washington and year-round residents across much of the rest of the state.

Food: While breeding, they eat insects and other arthropods; berries, fruits, and nuts are eaten mostly during the winter.

Nesting: April to May

Nest: Female builds a cup nest that is lined with dead leaves, moss, and grass.

Eggs: 1–6 light-blue eggs

Young: Chicks hatch with eyes closed and little down on their bodies after 12 days of incubation. They receive care from both parents. Chicks will leave the nest at around 15 days.

Predators: Raptors, pet cats, foxes, mink, raccoons, and snakes

Migration: Most of the population is made up of year-round residents with others as wintering residents.

Varied thrushes have long legs, short tails, rounded heads, and straight bills. They are usually seen standing on rocks or perching on horizontal limbs. They are slate-to-charcoal gray and blue on their back with burnt orange around their breast and belly. They have a black band on top of their breast and an orange stripe that extends from their eye down to the back of their head. Females are not as bright as males but have the same pattern.

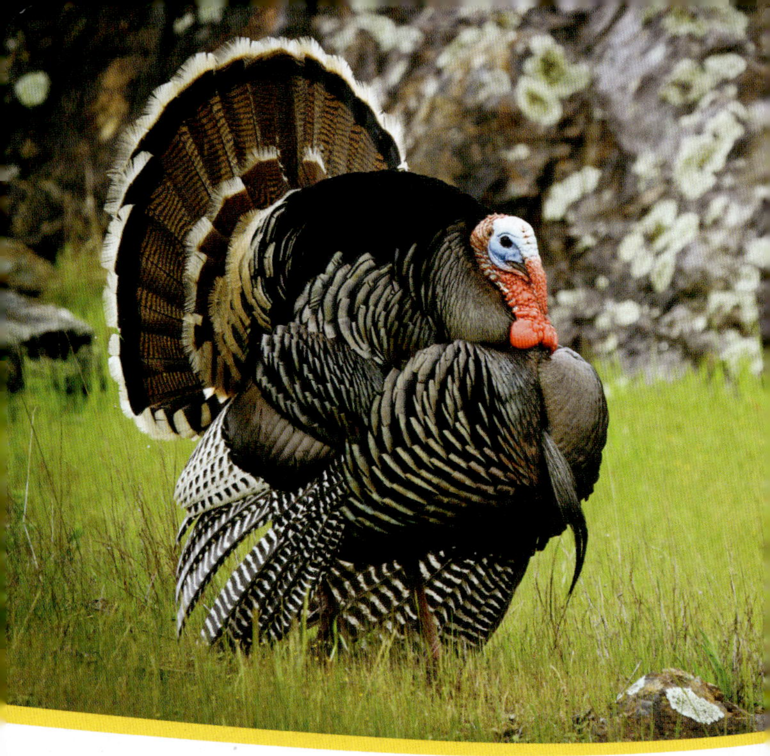

Did you know?

Turkeys sometimes fly at night, unlike most birds, and land in trees to roost. Turkeys have some interesting facial features; the red skin growth on a turkey's face above the beak is called a snood, while the growth under the beak is called a wattle. Wild turkeys can have more than 5,000 feathers.

Nest Type

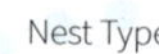

Most Active

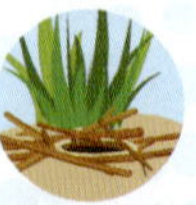

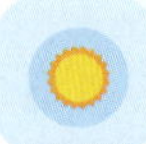

Wild Turkey

Meleagris gallopavo

Size: 3–4 feet long; wingspan of 5 feet; males weigh 16–25 pounds; females weigh 9–11 pounds

Habitat: Woodlands and grasslands

Range: They can be found in the eastern US and have been introduced in many western areas of the country. Though they are not native, they can be found throughout eastern Washington due to being introduced into the state in the early 20th century.

Food: Grain, snakes, frogs, insects, acorns, berries, and ferns

Nesting: April to September

Nest: The nest is built on the ground using leaves as bedding, in brush or near the base of trees or fallen logs.

Eggs: 10–12 tan eggs with very small reddish-brown spots

Young: Poults (young) hatch about a month after eggs are laid; they will flock with the mother for a year. When young are still unable to fly, the mom will stay on the ground with her poults to provide protection and warmth. When poults grow up, they are known as a hen if they are female, or a gobbler or tom if they are male.

Predators: Humans, foxes, raccoons, owls, eagles, and skunks

Migration: Turkeys do not migrate.

A wild turkey is a large bird that is dark brown and black with some iridescent feathers. Males will fan out their tail to attract a mate. When threatened, they will also fan out their tail and rush the predator, sometimes kicking and puncturing prey with the spurs on their feet.

Did you know?

The willow goldfinch, also known as the American goldfinch, helps restore habitats by spreading seeds. The goldfinch gets its color from a pigment called a carotenoid (say it, cuh-rot-en-oid) in the seeds it eats. It can even feed upside down by using its feet to bring seeds to its mouth.

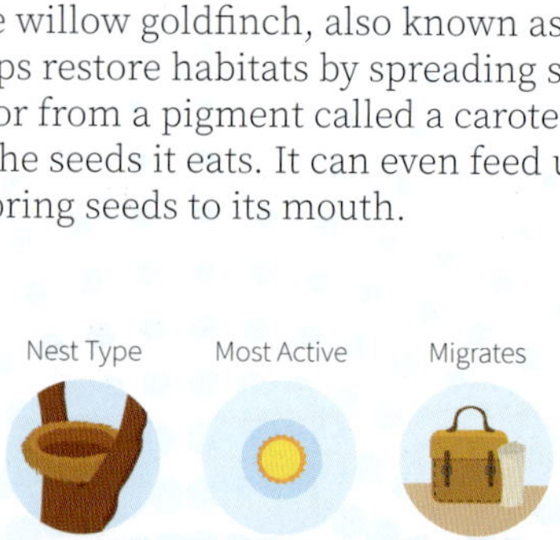

Nest Type

Most Active

Migrates

Willow (American) Goldfinch

Spinus tristis

Size: 4½–5 inches long; wingspan of 9 inches; weighs about half an ounce

Habitat: Grasslands, meadows, suburban areas, and wetlands

Range: In Washington, they are mostly year-round residents, with migratory visitors in the northwestern portion of the state. They can be found throughout much of the United States and southern Canada during various times of the year.

Food: Seeds of plants and trees; sometimes feeds on insects; loves thistle seeds at birdfeeders

Nesting: Goldfinches build a nest in late June.

Nest: Cup-shaped nests are built a couple of feet above-ground out of roots and plant fibers.

Eggs: 2–7 eggs with a bluish-white tint

Young: Young (chicks) hatch around 15 days after being laid; they hatch without feathers and weigh only a gram. Chicks learn to fly after around 11–15 days. Young become mature at around 11 months old.

Predators: Garter snakes, blue jays, American kestrels, and cats

Migration: Do not migrate much in Washington; in the north-western portion, birds will migrate north for breeding territories and south for wintering areas.

During the summer, willow goldfinch males are brightly colored with golden-yellow feathers and an orange beak. They have black wings with white wing bars. The crown (top) of the head is black. In winter, they molt, and the males look more like the females. Females are always greenish yellow with hints of yellow around the head.

Did you know?

Wood ducks will "mimic" a soccer player when a predator is near their young: they flop! Female wood ducks will fake a broken wing to lure predators away from their young. Wood duck chicks must jump from the nest after hatching to reach the water. They can jump 50 feet or more without hurting themselves.

Nest Type

Most Active

Migrates

Wood Duck

Aix sponsa

Size: 15–20 inches long; wingspan of 30 inches; weighs about a pound

Habitat: Swamps, woody ponds, and marshes

Range: In much of Washington, they are year-round residents, with populations along the Canadian border as breeding residents and the southern border as nonbreeding residents. They are also in the eastern US, southern Mexico, the Pacific Northwest, and on the West Coast.

Food: Fruits; nuts; and aquatic vegetation, especially duckweed, sedges, and grasses

Nesting: March to August

Nest: Wood ducks use hollow trees, abandoned woodpecker cavities, and human-made nesting boxes.

Eggs: 8–15 off-white eggs are laid once a year. Sometimes females will lay eggs in another female's nest; this process is called egg dumping.

Young: Eggs hatch about a month after being laid. Chicks will leave the nest after a day and fly within 8 weeks.

Predators: Raccoons, mink, fish, hawks, snapping turtles, owls, humans, and muskrats

Migration: Most do not migrate. Nonbreeding residents along the southern border migrate during the spring.

Wood duck males have a brightly colored crest (tuft of feathers) of iridescent (shimmering) green, red, and purple, with a mahogany (brown) upper breast area and tan bottom. Males also have red eyes. Females are brown to gray. Wood ducks have strong claws that enable them to climb up trees into cavities.

Did you know?

Kingsnakes have the widest distribution of any American snake. Kingsnakes get their name from their large size and the fact that they can eat snakes that are not only longer than they are but are also venomous. They have some immunity to (can't get hurt by) venom.

Most Active

Hibernates

California Mountain Kingsnake

Lampropeltis zonata

Size: 1½–3 feet long; weighs 3 pounds

Habitat: City areas, wetlands, grasslands, woodlands, forests, shrublands, and shortgrass prairies

Range: They can be found from Washington southward through Oregon into California, westward to Utah, and south into Arizona and Mexico. In Washington, they are restricted to two counties in the southern part of the state.

Food: They are carnivores that eat rodents, birds, lizards, snakes, and eggs.

Mating: Breed in spring once they come out of hibernation

Nest: No nest; they lay eggs under rocks, stumps, logs, or dying plant materials.

Eggs: 2–9 white eggs are laid at a time.

Young: Snakelets hatch 8–12 weeks after laying. They are 7–11 inches long. They stay hidden for about a week until they have shed their first skin. They reach adulthood 4–6 months after hatching.

Predators: Red-tailed hawks, pet cats, great horned owls, golden eagles, and other snakes, including other kingsnakes

California mountain kingsnakes are medium-size, nonvenomous snakes with smooth scales. They have black eyes with red, black, and cream-to-white bands going down their back from head to tail.

Did you know?
Coastal tailed frogs can't flick out their tongue in order to catch prey like other frogs because their tongue is not attached at the front of their mouth. Instead, they must actively catch their food.

Most Active

Coastal Tailed Frog

Ascaphus truei

Size: 1½–2 inches long; weighs around ¼ ounce

Habitat: Cold, clear, rocky, fast-moving streams

Range: They are found in western North America from coastal British Columbia to California. They can be found in western Washington from Chelan County westward to the Pacific Coast.

Food: Spiders, ticks, mites, snails, and various insects

Mating: Fall; males do not vocalize to find mate.

Nest: No nest; eggs are laid in flowing water attached to the undersides of rocks.

Eggs: Eggs are colorless and shaped like peas.

Young: Eggs will hatch in about 6 weeks. The first winter, they feed on their yolk sac; once their mouth is developed, they become more mobile and feed on algae. They remain tadpoles for 2–4 years.

Predators: Snakes, giant salamanders, trout and other fish, birds, insects, and mammals such as shrews

Coastal tailed frogs are small frogs with a large head and slender body. They have rough skin, vertical pupils, and eye stripes that flow from the snout to their shoulders. They are dark in color, often similar to the color of the stream bottom that they are in. Males have an appendage that looks like a tail, which is where they get their name. However, the "tail" is actually a reproductive organ. Unlike most frogs that fertilize their eggs after they are laid, this frog has internal fertilization. Males are smaller than females. Tadpoles are brown or black and many times have a white spot at the end of their tail.

Did you know?

Garter snakes are highly social. They will form groups with other snakes—and often other species—to overwinter together in a burrow or hole. When threatened by a predator or handled, they will sometimes musk or emit a foul-smelling, oily substance from their cloaca (butt). Garter snakes are able to eat the poisonous Pacific newt and may turn poisonous themselves for several weeks, making them deadly to any would-be predators.

Most Active

Hibernates

Common Garter Snake

Thamnophis sirtalis

Size: 18–26 inches long; weighs 5–5½ ounces

Habitat: Forests and forest edges, grasslands, and suburban areas

Range: They are found throughout much of the US and as far north as Canada. In Washington, they are found throughout the state.

Food: Frogs, snails, toads, salamanders, insects, fish, and worms

Mating: April or May

Nest: No nest; they will use natural cavities in the ground or abandoned burrows of small mammals.

Eggs: No eggs are laid. Garter snakes are born live in a litter of 8–20 snakes.

Young: Snakelets are 4½–9 inches long at birth; no parental care is given.

Predators: Crows, ravens, hawks, owls, raccoons, foxes, and squirrels

Common garter snakes are black with three yellow stripes running down their body on the back and sides. In eastern Washington, snakes have red blotches on the side of the body above stripes on their back. In western Washington, the blotches are not always present, and the color of the stripes varies from shades of yellow, green, blue, and even turquoise. They withstand winter by gathering in groups inside the burrows of rodents or under human-made structures, and they enter brumation, or a state of slowed body activity.

Did you know?

Cope's giant salamanders exhibit pedomorphosis (say it, pe-do-mor-pho-sis), which means they go through life without ever changing into terrestrial or land-living adults. They are the only giant salamanders that have been found in the Olympic Peninsula north of the Chehalis River.

Most Active

Hibernates

Cope's Giant Salamander

Dicamptodon copei

Size: 7½ inches long; weighs ¾–4 ounces

Habitat: Cold, clear, fast-flowing permanent streams in coniferous forests

Range: They are found in western Washington and northwestern Oregon.

Food: Carnivores, they eat small fish, tadpoles, eggs, insects, and other invertebrates.

Mating: Spring to fall

Nest: No nest; eggs are laid in water under rocks and logs.

Eggs: As many as 115 white eggs are laid one at a time. Females defend eggs until they hatch.

Young: Young hatch about 200 days after laying. Young very rarely metamorphose (transform) into a terrestrial adult (able to live on land).

Predators: River otters, weasels, snakes, shrews, fish, and mink

Medium-size salamanders, Cope's giant salamanders have a rounded snout, bushy gills, and a flattened tail. Their color is brown to marbled gold. Adults that go through metamorphosis are brown with areas of blue, and they have a narrow head.

Did you know?

The northern and southern alligator lizards are the only lizards in Washington with square scales. They can only be found on the West Coast.

Most Active

Hibernates

Northern (NAL)/Southern (SAL) Alligator Lizard

Elgaris coerulea/Elgaria multicarinata

Size: NAL: 10 inches long; weighs 1 ounce. SAL: 12 inches long; weighs 1 ounce

Habitat: NAL: rocky areas, grassy openings, bushy areas, and edges of lakes. SAL: rocky, grassy areas; brushy openings; forests; creeks; and streams

Range: NAL: can be found from Canada to northern California; in Washington, they can be found almost statewide. SAL: has a similar range that extends down into Mexico; found in a limited range in southern Washington.

Food: Insects, spiders, snails, scorpions, millipedes, tadpoles, lizards, and sometimes birds and eggs

Mating: April to May

Nest: NAL: no nest; SAL: rock crevice, decaying wood or plant material, and rodent burrows

Eggs: NAL: no eggs; give birth to live young. SAL: 5–20 eggs

Young: NAL: 2–15 young are born at a time. They are independent at birth and are reproductively mature at around 2 years. SAL: Eggs hatch about 11 weeks after laying. They reach reproductive maturity at 19 months.

Predators: Snakes including rubber boas and racers, shrikes, red-tailed hawks, and cats

Northern alligator lizards are grayish brown to brown. They have brown eyes and a light-colored belly with dark spots on the side. Males have a larger head than females. Young are bronze or copper with a big bronze stripe on the back. Southern alligator lizards are large and brown to grayish in color. They have yellow or golden eyes. The underside is light without dark spots on the sides. Both have rough scales, short legs, and a long tail.

Did you know?
Turtles are the oldest living reptiles! The first turtle is believed to have shown up over 215 million years ago. Western pond turtles are one of two freshwater turtles native to Washington as well as the entire Pacific Northwest.

Most Active

Hibernates

Northwestern Pond Turtle

Actinemys marmorata

Size: 3½–9 inches long; weighs 1–2½ pounds

Habitat: Slow-moving rivers, streams, and creeks; small lakes and ponds; marshes; human-made bodies of water

Range: Although once found from Canada down into California, populations presently reside in southern Washington, Oregon, California, and Mexico. In Washington, they are limited to two sites in the Columbia River and George Strait areas of the state.

Food: Fish, tadpoles, worms, frogs, eggs of frogs and salamanders, crayfish, and carrion (dead animals)

Mating: Late April or early May

Nest: Nests are made on sandy banks or in fields that get a lot of sun and are close to water.

Eggs: 1–13 eggs

Young: Young are only 1 inch long and independent at hatching. Sex of hatchlings is based on the temperature of eggs: 85 degrees and below produce males, and 86 degrees and above produce females. They reach reproductive maturity at around 8–10 years old.

Predators: Fish, bullfrogs, snakes, wading birds, and mammals

Western pond turtles are medium-size turtles with a yellow belly (plastron). Their skin is yellow or cream and spotted with blotches of black. Males are smaller than females, have paler chins and throats, and have thicker tails. Females have smaller heads, flatter plastrons, darker chins, and a carapace (top of shell) that is more dome shaped.

Did you know?

The Oregon spotted frog has a clicking call that is so low pitched that you have to be within 15–16 feet to hear it. During winter, these frogs will hibernate in a water source under a foot of mud.

Most Active

Hibernates

Oregon Spotted Frog

Rana pretiosa

Size: 2–4 inches long; weighs 1–3 ounces

Habitat: Streams, forest, grasslands, and wetlands

Range: They can be found in western Washington as well as western and central Oregon.

Food: Carnivores, they feed on insects and other invertebrates.

Mating: Early spring

Nest: No nest; egg masses are laid in water.

Eggs: Egg masses are typically communally laid, resulting in clustered groups of 2 to more than 100.

Young: Young hatch into tadpoles within a month and will metamorphize (change into frogs) in 100 days. They reach reproductive maturity at 2–3 years.

Predators: Garter snakes, beetles, American bullfrogs, fish, birds, cats, rough-skinned newts, raccoons, and giant water bugs

Medium-size frogs, Oregon spotted frogs are brick red to olive brown with black spots that have light centers. They have red-to-orange-red undersides and legs. They have almost fully webbed feet.

Did you know?

The Pacific chorus frog is the most abundant species of frog on the West Coast of North America and the most common frog species in Washington. Pacific chorus frogs are the only frogs that actually say, "Ribbit," which is the sound used for most toys and materials that need a frog noise.

Most Active

Hibernates

Pacific Chorus Frog

Pseudacris regilla

Size: 1–2 inches long; weighs about ⅖ gram

Habitat: Meadows, open fields, lakes, forest edges, and ponds

Range: There are strong populations from Canada southward into Washington and Montana. They can be found as far south as Baja California, Mexico, and westward to Nevada. In Washington, they can be found throughout the state.

Food: Spiders, worms, insects, and other invertebrates

Mating: November to July; mating occurs in water.

Nest: No nest is constructed; female will lay eggs in permanent shallow bodies of water, attached to vegetation just below the surface.

Eggs: Eggs are laid in loose clusters of about 8–70 eggs (as many as 750 eggs) that are usually attached to vegetation, though they may sometimes be laid on the sediment or ground near water sources. The clusters are about 1–1½ inches in diameter.

Young: Tadpoles hatch 2–5 weeks after laying and will metamorphize (turn into frogs) within 2½ months. They are brown to olive in color with black speckles and a white underside.

Predators: Snakes, birds, giant water bugs, raccoons, and fish

Female Pacific chorus frogs are larger than males. Both sexes have circular discs on their toes called toe pads. Chorus frogs are green or brown on their back (dorsal side); they also come in various shades of gray and are white on the underside. They have a dark mask that runs from the nose through both eyes, and they have a Y-shaped mark in between their eyes.

Did you know?

The rubber boa's common name comes from the fact that its skin is loose and wrinkly, making it look like it is made of rubber. This snake is the smallest member of the Boa family. When threatened, rubber boas coil their body with their head down and stick out their tail, which is similar in appearance to their head, fooling predators into biting it.

Most Active

Hibernates

Rubber Boa

Charina bottae

Size: 21–26 inches long; weighs 2½–7 ounces

Habitat: Prairies, shrub-steppe, grasslands, forests, deserts, and foothills

Range: They can be found from southern California northward to Canada, and as far west as Utah and Wyoming. They are found throughout much of the state of Washington

Food: Eats small mammals, birds, salamanders, lizards, snakes, and possibly frogs

Mating: Spring

Nest: No nest is made; they give live birth usually in underground burrows, under vegetation, or in rock crevices.

Eggs: No eggs are laid; they birth live young.

Young: Young are born pink, and their skin darkens as they age. They are about 7–11 inches long. They reach maturity at around 2–3 years.

Predators: Bobcats, pet cats, birds of prey, coyotes, ravens, and raccoons

Rubber boas are small, thick-bodied snakes with small scales. They have a thumb-like head and similarly shaped tail. They come in various colors from rosy pink to brown, gray, and charcoal. Their underside is yellow or cream. The females are larger than the males.

Did you know?

This turtle can breathe through its butt! Known technically as "cloacal breathing," this adaptation enables them to hibernate underwater. They still need to breathe, of course, but their body slows down so much that the oxygen they absorb through blood vessels in their behind is enough for them to survive.

Most Active

Hibernates

Western Painted Turtle

Chrysemys picta bellii

Size: Male is 7 inches long; female is 10–12 inches long; males weigh 10–12 ounces; females weigh around 18 ounces.

Habitat: Marshes, small lakes, ponds, wetlands, and swamps

Range: They can be found from eastern Canada all the way to the Pacific Northwest. They are found across the state of Washington; in some areas like the Puget Sound region, they have also been introduced.

Food: Fish, tadpoles, dead animals, crayfish, and insects

Mating: Late spring and fall

Nest: The turtle digs a hole in soil/sand adjacent to a body of water.

Eggs: 8–9 oval eggs; eggs have a soft shell.

Young: Hatchlings emerge from the nest between August and September; young rely on the egg yolk for food for the first few days of life. In their first year, they can double their size. Females reach reproductive maturity around 11–16 years, while males reach it around 7–9 years.

Predators: Snapping turtles, raccoons, opossums, foxes, mink, skunks, cats, humans, and fish

The painted turtle's skin is black with two yellow stripes that line the head. The carapace (top shell) is black, while the plastron (bottom shell) is yellow or even red sometimes. Painted turtles can be found basking in groups. Painted turtles can live over 35 years in the wild, although most do not live to be that old.

Did you know?

The western rattlesnake is the only venomous snake found in Washington. It will rattle the base of its tail to warn would-be predators; it can move its rattle back and forth 60 or more times per second. Do not go near one or try to pick one up! Instead, leave it alone so it can help people by munching on rodents and other pests. They are "pit vipers," which are snakes that have a special body part that helps them "see" heat.

Most Active

Hibernates

Safety Note: These snakes are venomous (toxic). If you see one, observe or admire it from a distance.

Western Rattlesnake

Crotalus oreganus

Size: 3–3½ feet long; weighs 2 pounds

Habitat: Rocky hillside, grassy plains, coastal areas, deserts, and forests

Range: They are found in the western states of the US and the northern half of Mexico. They are found in the eastern portion of Washington that falls east of the Cascade Mountains.

Food: Carnivores, they feed on mice, rats, rabbits, gophers, ground-dwelling birds, bird eggs, and lizards.

Mating: Spring following hibernation

Nest: Gives live birth in a burrow or hollow log

Eggs: No eggs; young are birthed live.

Young: 10–12 young are born at a time. Young will stay with the mother for a few hours and scatter to find shelter and live independently. Young have fangs and venom at birth. They reach reproductive maturity at around 3 years.

Predators: Coyotes, eagles, hawks, foxes, kingsnakes, bobcats, and roadrunners

The western rattlesnake is a thick snake with an arrow- or triangle-shaped head. Two diagonal lines that are dark colored run from the snout through the eyes, as if it is wearing a mask. It has dark, diamond-shaped patterns that run down its olive-brown or tan body. These snakes have brown spots bordered by black rings, on top of white blotches running down their back. The spots on the back turn to bands as they reach the tail. Their body color ranges in different shades of brown to olive. Their underside is pale yellow.

Did you know?

The western skink is the only lizard in Washington that has a bright-blue tail. If threatened or caught by a predator, skinks will drop their tail as a distraction and escape as the predator focuses on the wiggling tail. They are great at burrowing and sometimes construct burrows that are multiple times longer than they are.

Most Active

Hibernates

Western Skink

Plestiodon skiltonianus

Size: 4–8¼ inches long; weighs ⅕–¼ ounce

Habitat: Rocky areas, grasslands, chaparral, shrublands, woodlands, and forests

Range: They are found in the western US and Canada. In Washington, they are found in the eastern part of the state.

Food: Carnivores, they feed on crickets, beetles, moths, and other insects, as well as spiders and other arthropods.

Mating: Spring

Nest: Eggs are laid during June and July in cavities or chambers or under a rock in the ground made by the female.

Eggs: 2–6 eggs are laid per clutch.

Young: Eggs hatch in the summer. Young are reproductively mature at 2 years, though most will not reproduce until 3 years.

Predators: Snakes, crows, loggerhead strikes, pet cats, shrews, and birds of prey

Western skinks have a long, slender body with a tail that is about the same size as the body. The body is covered in rounded scales that are smooth and shiny. They are brown or black on their back or dorsal side. They have cream-to-golden-yellow stripes that run from the tip of the nose down to the tail and a broad, brown stripe that runs between the two smaller ones. The underside is gray to cream with a green-and-blue hue. During breeding season, males have an orangish-red color on the sides of their head and chin. Young skinks have a blue tail.

Did you know?

The western tiger salamander can grow up to 14 inches long and live over 20 years! Western tiger salamanders migrate to their birthplace to breed. Tiger salamanders have a hidden weapon! They produce a poisonous toxin that is secreted or released from two glands in their tail. This toxin makes them taste bad to predators and allows them to escape.

Most Active

Hibernates

Western Tiger Salamander

Ambystoma mavortium

Size: 7–14 inches long; weighs 4½ ounces

Habitat: Woodlands, marshes, and meadows; they spend most of their time underground in burrows.

Range: Populations are found in the western United States. In Washington, they are found in the western portion of the state in the Colombia Plateau.

Food: They are carnivores that eat insects, frogs, worms, and snails.

Mating: Tiger salamanders leave their burrows to find standing bodies of freshwater. They breed in late winter and early spring after the ground has thawed.

Nest: No nest, but eggs are joined together into one group in a jelly-like sack called an egg mass. An egg mass is attached to grass, leaves, and other plant material at the bottom of a pond.

Eggs: There are 20–100 eggs or more in an egg mass.

Young: Eggs hatch after 2 weeks, and the young are fully aquatic with external gills. Limbs develop shortly after hatching; within 3 months, the young are fully grown but will hang around in a vernal pool. Individuals living in permanent ponds can take up to 6 months to fully develop.

Predators: Young are preyed upon by diving beetles, fish, turtles, and herons. Adults are preyed upon by snakes, owls, and badgers.

Western tiger salamanders have thick black, brown, or grayish bodies with uneven spots of yellow, tan, brown, or green along the head and body. The underside is usually a variation of yellow. Males are usually larger and thicker than females.

Did you know?

While they are called racers, they can only move about 4 miles per hour. When threatened, they will musk themselves (secrete a rancid-smelling fluid) from their scent gland. Juvenile racers are sometimes confused with another type of snake because they look so different from the adults.

Most Active

Hibernates

Western Yellow-bellied Racer

Coluber constrictor mormon

Size: 3–6 feet long; weighs about 1¼ pounds

Habitat: Marshes, bogs, woodlands, open fields, thickets, lake edges, and meadows

Range: They are found from southern Canada down through Washington, as far south as California, and as far west as Colorado. In Washington, they are found in the eastern part of the state.

Food: Small mammals, nestling birds, eggs, spiders, insects, small turtles, snakes, and frogs

Mating: Early summer

Nest: Eggs are laid in cavities like animal burrows and logs that are decaying.

Eggs: 3–11 white oval eggs

Young: Eggs hatch in August or early September. Snakelets are 8–11 inches long. They are independent at birth. Males reach reproductive maturity at around 2 years, females at 2 or 3 years.

Predators: Other snakes, pet cats, hawks and eagles, foxes, and coyotes

Western yellow-bellied racers are narrow-headed slender snakes with big eyes, a dark body, and a light-colored underside. Juveniles have blotching on their back.

Glossary

Adaptation—An animal's physical (outward) or behavioral (inward) adjustment to changes in the environment.

Amphibian—A small animal with a backbone, moist skin, and no scales. Most amphibians start out as an egg, live at least part of their life in water, and finish life as a land dweller.

Biome—A part or region of Earth that has a particular type of climate and animals and plants that adapted to live in the area.

Bird—A group of animals that all have two legs and feet, a beak, feathers, and wings; while not all birds fly, all birds lay eggs.

Brood—A group of young birds that hatch at the same time and with the same mother.

Carnivore—An animal that primarily eats other animals.

Clutch—The number of eggs an animal lays during one nesting period; an animal can lay more than one clutch each season.

Crepuscular—The hours before sunset or just after sunrise; some animals have adapted to be most active during these low-light times.

Diurnal—During the day; many animals are most active during the daytime.

Ecosystem—A group of animals and plants that interact with each other and the physical area that they live in.

Evolution—A process of change in a species or a group of animals that are all the same kind; evolution happens over several generations or in a group of animals living around the same time; evolution happens through adaptation, or physical and biological changes to better fit the environment over time.

Fledgling—A baby bird that has developed flight feathers and has left the nest.

Gestation—The length of time a developing mammal is carried in its mother's womb.

Herbivore—An animal that primarily eats plants.

Hibernation—A survival strategy or process where animals "slow down" and go into a long period of reduced activity to survive winter or seasonal changes; during hibernation, activities like feeding, breathing, and converting food to energy all stop.

Insectivore—An animal whose diet consists of insects.

Incubate—When a bird warms eggs by sitting on them.

Invasive—A non-native animal that outcompetes native animals in a particular area, harming the environment.

Mammal—An air-breathing, warm-blooded, fur- or hair-covered animal with a backbone. All mammals produce milk and usually give birth to live young.

Migration—When animals move from one area to another. Migration usually occurs seasonally, but it can also happen due to biological processes, such as breeding.

Molt—When animals shed or drop their skin, feathers, or shell.

Nocturnal—At night; many animals are most active at night.

Piscivore—An animal that eats mainly fish.

Predator—An animal that hunts (and eats) other animals.

Raptor—A group of birds that all have a curved beak and sharp talons; they hunt or feed on other animals. Also known as a bird of prey.

Reptile—An egg-laying, air-breathing, cold-blooded animal that has a backbone and skin made of scales, which crawls on its belly or uses stubby legs to get around.

Scat—The waste product that animals release from their bodies; another word for it is poop or droppings.

Talon—The claw on the feet seen on raptors and birds of prey.

Torpor—A form of hibernation in which an animal slows down its breathing and heart rate; torpor ranges from a few hours at a time to a whole day. Torpor does not involve a deep sleep.

Checklist

Mammals

- [ ] American Badger
- [ ] American Beaver
- [ ] Black Bear
- [ ] Bobcat
- [ ] California Gray Whale
- [ ] California Sea Lion
- [ ] Coyote
- [ ] Gray Wolf
- [ ] Grizzly Bear
- [ ] Harbor Seal
- [ ] Hoary Bat
- [ ] Humpback Whale
- [ ] Mountain Goat
- [ ] Mountain Lion
- [ ] Mule Deer
- [ ] Northern American Porcupine
- [ ] Northern Raccoon
- [ ] Northern River Otter
- [ ] Northern Sea Otter
- [ ] Olympic Marmot
- [ ] Orca
- [ ] Pika
- [ ] Roosevelt/ Rocky Mountain Elk
- [ ] Snowshoe Hare
- [ ] Striped Skunk

Birds

- [ ] American Robin
- [ ] Anna's Hummingbird
- [ ] Bald Eagle
- [ ] Black-capped Chickadee
- [ ] Bufflehead
- [ ] Canada Goose
- [ ] Common Raven/ American Crow
- [ ] Dark-eyed Junco
- [ ] Double-crested/ Pelagic Cormorant
- [ ] Golden Eagle
- [ ] Great Blue Heron
- [ ] Great Gray Owl
- [ ] Great Horned Owl
- [ ] Hairy/Downy Woodpecker
- [ ] Northern Flicker
- [ ] Osprey
- [ ] Red-tailed Hawk

- [ ] Red-winged Blackbird
- [ ] Steller's Jay
- [ ] Tufted Puffin
- [ ] Varied Thrush
- [ ] Wild Turkey
- [ ] Willow (American) Goldfinch
- [ ] Wood Duck

Reptiles and Amphibians

- [ ] California Mountain Kingsnake
- [ ] Coastal Tailed Frog
- [ ] Common Garter Snake
- [ ] Cope's Giant Salamander
- [ ] Northern/Southern Alligator Lizard
- [ ] Northwestern Pond Turtle
- [ ] Oregon Spotted Frog
- [ ] Pacific Chorus Frog
- [ ] Rubber Boa
- [ ] Western Painted Turtle
- [ ] Western Rattlesnake
- [ ] Western Skink
- [ ] Western Tiger Salamander
- [ ] Western Yellow-bellied Racer

The Art of Conservation®

Featuring two signature programs, The Songbird Art Contest™ and The Fish Art Contest®, the Art of Conservation programs celebrate the arts as a cornerstone to conservation. To enter, youth artists create an original hand-drawn illustration and written essay, story, or poem synthesizing what they have learned. The contests are FREE to enter and open to students in K-12. For program updates, rules, guidelines, and entry forms, visit: www.TheArtofConservation.org

The Fish Art Contest® introduces youth to the wonders of fish, the joy of fishing, and the importance of aquatic conservation. The Fish Art Contest uses art, science, and creative writing to foster connections to the outdoors and inspire the next generation of stewards. Participants are encouraged to use the Fish On! lesson plan, then submit an original, handmade piece of artwork to compete for prizes and international recognition.

The Songbird Art Contest® explores the wonders and species diversity of North American songbirds. Raising awareness and educating the public on bird conservation, the Songbird program builds stewardship, encourages outdoors participation, and promotes the discovery of nature.

Photo Credits

Stewart Ragan: 144; **Jonathan Norberg:** 23t, 31t

Silhouettes and tracks by Anthony Hertzel unless otherwise noted.
s=silhouette, t=animal track(s)

This image is used under Attribution 2.0 Generic (CC BY 2.0) license, which can be found at https://creativecommons.org/licenses/by/2.0/: **Forest and Kim Starr:** 33, no modifications, original image at www.flickr.com/photos/starr-environmental/24755918320

This image is used under Attribution 4.0 International (CC BY 4.0) license, which can be found at https://creativecommons.org/licenses/by/4.0/: **66dodge:** 123, no modifications, original image at www.inaturalist.org/photos/1285257; **Jeremiah Degenhardt:** 133, no modifications, original image at www.inaturalist.org/photos/12559978; **James M. Maley:** 130, no modifications, original image at www.inaturalist.org/photos/282336828; **Melissa McMasters:** 122, no modifications, original image at www.inaturalist.org/photos/147881226; **Ken-ichi Ueda:** 112, no modifications, original image at www.inaturalist.org/photos/59901,131, no modifications, original image at www.inaturalist.org/photos/3402248, 132, no modifications, original image at www.inaturalist.org/photos/245455806; **Colton Veltkamp:** 113, no modifications, original image at www.inaturalist.org/photos/377953398

All images used under license from Shutterstock.com:
John L. Absher: 92; **ace03:** footer burst; **Airin.dizain:** 44s; **Harvest Aksell:** 23; **Muhammad Alfatih 05:** 18s; **Alpha C:** 26s, 42s, 60s; **ANDRES MENA PHOTOS:** 80; **Victor Arita:** 19; **Agnieszka Bacal:** 76; **Raul Baena:** 107; **Michael Benard:** 119, 137; **Randy Bjorklund:** 51, 114; **Roman Bjuty:** 31, **BLAKE MCKENNA PHOTOGRAPHY:** 99; **Todd Boland:** 105; **Steve Byland:** 69; **Mark Castiglia:** 17; **Mike Chan:** 40, 41; **yongsheng chen:** 86; **Ivan Chistyakov:** 8 (apple); **Jesus Cobaleda:** 22; **Mircea Costina:** 78 (double-crested cormorant); **Creeping Things:** 121; **Jim Cumming:** 26, 74 (American crow); **Chase D'animulls:** 28, 70; **Danita Delimont:** 8 (western hemlock); **dimostudio:** 98; **dmvphotos:** 48; **DnDavis:** 87; **Dominate Studio:** 14s; **Ian Duffield:** 94; **Earth theater:** 34; **Elementspace:** 125; **Nico Faramaz:** 35; **Deborah Ferrin:** 25; **Frank Fichtmueller:** 15; **FloridaStock:** 66; **FotoRequest:** 8 (willow goldfinch), 89, 106; **Bildagentur Zoonar GmbH:** 61; **iliuta goean:** 81; **grandbrothers:** 77; **Elliotte Rusty Harold:** 46; **Harry Collins Photography:** 108; **Ayman Haykal:** 63; **Hill Photography:** 36; **Karen Hogan:** 88; **HWall:** 8 (Pacific chorus frog), 116, 117; **JamesChen:** 90; **JCWaller:** 32; **Matt Jeppson:** 110, 111, 115, 126, 127, 134, 136; **Anne Katherine Jones:** 71; **Paul Jones Jr:** 72; **Joseph Scott Photography:** 83; **Tory Kallman:** 18, 21, 104; **Viktoria Karpunina:** 58s; **KBel:** 20s, 34s; **Janet M Kessler:** 62; **Kilmer Media:** 8 (state seal); **Chris Klonowski:** 75; **Rich Koele:** 8 (Columbian mammoth); **Krumpelman Photography:** 96; **Piotr Krzeslak:** 74 (common raven), 84; **Geoffrey Kuchera:** 60; **Holly Kuchera:** 27, 42; **Dennis Laughlin:** 12, 124; **Sean Lema:** 8 (steelhead trout); **Cliff LeSergent:** 64, 65; **David A Litman:** 79, 118; **L-N:** 45; **mamita:** 24s; **Don Mammoser:** 95; **Karl R. Martin:** 82; **Kazakova Maryia:** 11 (ground nest); **Martin Mecnarowski:** 93; **Mekong Photography:** 120; **Elly Miller:** 67; **Miloje:** background/inset burst; **Christian Musat:** 14; **nialat:** 47; **Paul Reeves Photography:** 44; **Pavel K:** 12t; **Rita Petcu:** 43; **Peter Turner Photography:** 8 (evergreen state); **pichayasri:** 11 (platform nest), 11 (suitcase); **PomInPerth:** 85; **predragilievski:** 56s; **Martin Prochazkacz:** 20; **Tom Reichner:** 55, 59; **reptiles4all:** 129, 135; **Robert Harding Video:** 101; **RRichard29:** 58; **RubenAlfaro:** 50; **Ryguyryguy74:** 13; **George Schmiesing:** 97; **David M Schultz:** 49; **SCStock:** 16; **Waldemar Manfred Seehagen:** 38; **ShayneKayePhoto:** 54, 103; **SiK Imagery:** 39; **slowmotiongli:** 53; **Rowdy Soetisna:** 102; **SofiaV:** 11 (cavity nest); **sreewing:** 56t; **Rostislav Stach:** 56, 57; **Stock_Ninja:** 48s; **stopkin:** 16s; **Marek R. Swadzba:** 73; **T_DubOv:** 11 (cup nest); **tab62:** 8 (coast rhododendron); **Jennifer Tepp:** 8 (green darner dragonfly); **Thomas Torget:** 24; **toseeg:** 29; **Tereza Tothova:** 8 (orca); **tryton2011:** 100; **Vladimir Turkenich:** 52; **vagabond54:** 78 (pelagic cormorant); **Vector412:** 30s; **vectoric:** 10 (basketball); **Viktorya170377:** 22s, 52s; **Wirestock Creators:** 91; **Brian A Wolf:** 37; **ya_mayka:** 46s; **yhelfman:** 30; **yvontrep:** 68; **Peter K. Ziminski:** 128; **Oral Zirek:** 109

About the Author

Alex Troutman is a wildlife biologist, birder, nature enthusiast, and science communicator from Austell, Georgia. He has a passion for sharing the wonders of nature and introducing the younger generation to the outdoors. He holds both a bachelor's degree and a master's degree in biology from Georgia Southern University (the Real GSU), with a focus in conservation. Alex knows what it feels like not to see individuals who look like you (or come from a similar background) doing the things you enjoy or working in the career that you aspire to be in. He makes a point to be that representation for the younger generation, ensuring that kids have exposure to the careers they are interested in and to the diverse scientists working in those careers.

Alex is the co-organizer of several Black in X weeks, including Black Birders Week, Black Mammalogists Week, and Black in Marine Science Week. This movement encourages diversity in nature, the celebration of Black individual scientists, awareness of Black nature enthusiasts, and diversity in STEAM fields.

ABOUT ADVENTUREKEEN

We are an independent nature and outdoor activity publisher. Our founding dates back more than 40 years, guided then and now by our love of being in the woods and on the water, by our passion for reading and books, and by the sense of wonder and discovery made possible by spending time recreating outdoors in beautiful places. It is our mission to share that wonder and fun with our readers, especially with those who haven't yet experienced all the physical and mental health benefits that nature and outdoor activity can bring. #bewellbeoutdoors